Scalable Innovation

A Guide for Inventors, Entrepreneurs, and IP Professionals

Scalable Innovation

A Guide for Inventors, Entrepreneurs, and IP Professionals

EUGENE SHTEYN
MAX SHTEIN

CRC Press is an imprint of the
Taylor & Francis Group, an **informa** business

CRC Press
Taylor & Francis Group
6000 Broken Sound Parkway NW, Suite 300
Boca Raton, FL 33487-2742

Printed on acid-free paper
Version Date: 20130412

International Standard Book Number-13: 978-1-4665-9097-7 (Paperback)

Library of Congress Cataloging-in-Publication Data

Shteyn, Eugene, 1961-
Scalable innovation : a guide for investors, entrepreneurs, and IP professionals / Eugene Shteyn and Max Shtein.
pages cm
"A CRC title, part of the Taylor & Francis imprint, a member of the Taylor & Francis Group, the academic division of T&F Informa plc."
Includes bibliographical references and index.
ISBN 978-1-4665-9097-7 (paperback : acid-free paper)
1. Inventions--Handbooks, manuals, etc. 2. Technological innovations--Handbooks, manuals, etc. 3. Inventions--Economic aspects--Handbooks, manuals, etc. 4. Technological innovations--Economic aspects--Handbooks, manuals, etc. I. Shtein, Max, 1976- II. Title.

T339.S57 2013
600--dc23 2013000866

Visit the Taylor & Francis Web site at
http://www.taylorandfrancis.com

and the CRC Press Web site at
http://www.crcpress.com

Contents

SECTION III System Evolution and Innovation Timing

Foreword

Today, most people agree that innovation is a major driving force behind economic growth and general improvement in human well-being. Research shows that in countries with social institutions that promote innovation, people not only live longer, but also enjoy high standards of living [1]. Governments and private companies pour money into research and development (R&D). In 2013, US R&D spending is expected to reach $423.7 billion, with global R&D increasing 3.7% to nearly $1.5 trillion [2]. Some US higher education institutions have successfully moved beyond the traditional educational and scientific research role to the model of a modern entrepreneurial university. For example, companies founded by Stanford University entrepreneurs generate world revenues of $2.7 trillion annually [3]. Similar estimates for MIT show a wealth-generation level of $2 trillion in worldwide sales.

Nevertheless, there's strong evidence that the engine of innovation is faltering. In the book *The Great Stagnation*, economist Tyler Cowen argues that since the 1970s, technological innovation has failed to deliver significant real growth in US median income and the last three economic recoveries have not led to job growth. He writes, "[D]uring the last forty years, that low-hanging fruit started disappearing, and we started pretending it was still there. We have failed to recognize that we are at a technological plateau and the trees are more bare than we would like to think" [4].

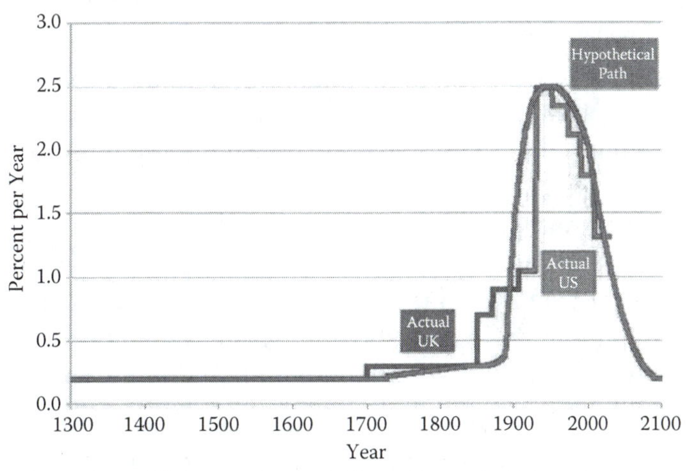

Growth in real GDP per capita, 1300-2100, with actual and hypothetical paths. *Source*: Robert J. Gordon. Is U.S. economic growth over? Faltering innovation confronts the six headwinds. 2012. NBER Working Paper 18315. [5]

In a 2012 National Bureau of Economic Research paper, Robert J. Gordon shows a slowdown in innovation-related growth in the United States. He identifies powerful headwinds and suggests that the rapid growth created by the industrial revolutions of the past might have been a singular chain of events [5].

In a November 3, 2012, *New York Times* article, Clayton Christensen, the author of a seminal book *The Innovator's Dilemma,* discusses possible reasons for the diminishing returns on innovation and argues that instead of investing in potential breakthroughs, we put too much money into incremental, low-risk technological improvements. Furthermore, he argues that our education system is making the problem worse by not teaching enough of "the educational skills necessary to start companies that focus on empowering innovations" [6].

The paths of innovation are risky and filled with failures. According to some estimates, three thousand raw ideas result in just one successful product. Even in Silicon Valley, the hotbed of successful technology entrepreneurship, less than 1 out 20 startups make it beyond a few years [7].

We believe that this situation is a terrible waste of creative talent and it should be changed. It is time to rethink the way we do innovation. Or, in the words of Geoffrey West in his TED talk on scalability of biology and economy (titled "The Surprising Math of Cities and Corporations"), "we are not only on a treadmill that's going faster, but we have to change the treadmill faster and faster" [8].

In this book we develop novel tools for inventors and entrepreneurs, and propose systematic methods to improve the quality of idea generation and the timing of innovation commercialization. We base our approach on the advances in cognitive science that show how common psychological biases lead to systematic mistakes in understanding and solving problems [9], and we take into account recent research into scalable systems, spanning biological organisms, corporations, cities, and so on [8]. Finally, we apply methods and models developed for systematic problem solving that go well beyond traditional brainstorming techniques.

In a typical process of problem solving at school (kindergarten to university tests), or on the job, the term *problem* more or less means a puzzle with a predetermined answer. In this sense, solving a problem means completing a jigsaw puzzle, and a person's ranking in the class or work place depends on how quickly they solve puzzles. In other words, a problem is perceived as any deviation from an established process or state of affairs, and returning a system back to its previous *normal* state constitutes solving the problem or completing the jigsaw puzzle.*

In contrast, some of the biggest problems facing today's inventors and entrepreneurs require solutions that do away with the established state of affairs. They require breakthrough innovations that make existing approaches obsolete. Thus, in contrast to the traditional model, we set out to identify problems whose solutions create new systems, establish new processes, and set new agendas. Doing so successfully is likely to trigger new, follow-up problems, which then could be solved using a variety of approaches. This entire process of creating a world of opportunities constitutes scalable innovation.

The development of practical innovation skills is hampered significantly by typical narratives of innovation stories, where innovation is presented as *creative destruction* or as *disruptive*. These broadly accepted narratives implicitly convey the point of view of established industry players whose businesses are being destroyed or disrupted by a new startup. Since obsolescence of the old is just a side effect of a successful innovation, we believe it is far more productive for an inventor or technology entrepreneur to spend most of his or her attention on finding new paths and creating new worlds, instead of focusing on disrupting any existing institution or practice. This subtle shift of focus helps develop a fresh perspective on notions people commonly perceive as given and immutable.

For example, innovation forecasts follow the standard calendar time, that is, measured in months, quarters, Moon phases, and Earth's rotation cycles. But unless the proposed solution deals with space travel, agriculture, or other season-related issues, early-stage innovation has little to do with astronomy. Instead, the approach introduced in this book treats the issue of timing as a relationship between elements of technology and business models, and their constraints. We focus on improving our readers' understanding of innovation as a systematic process, where problems and solutions don't come up randomly, but rather through a series of emerging patterns of interactions between system elements. Our goal is to help readers discover quickly these patterns and take advantage of them to create breakthrough opportunities.

---

* Per Sir Ken Robinson, the post-Industrial Revolution education system is geared toward mass market education, which is in turn geared toward upkeep of existing systems. Thinking about the mass-market educational system as an economic enterprise that must be scalable, we can understand the perceived need to "measure" learning outcomes very quickly and to rank people accordingly. Sir Ken Robinson, PhD, is an internationally recognized leader in the development of education, creativity, and innovation. http://Sirkenrobinson.com/skr/

In the Prologue, we show how common misconceptions—proliferated through traditional education, work, and popular culture—hurt our ability to identify and take advantage of breakthroughs. The key takeaway messages from that part of the book are as follows

1. We can improve our chances for successful innovation by rejecting, instead of accepting, the notion that "everything is a trade-off and there's no such thing as a free lunch." Research shows that a trade-off is usually an indication of a constraint everybody takes for granted. We provide real-life examples of how successful innovators create opportunities for scalable innovations specifically by breaking a broadly accepted trade-off.
2. Despite carrying similar-sounding names, *invention* and *innovation* constitute different activities. The former is a process for generating ideas. The latter is a process of making ideas *work* for the largest number of people possible. For an idea to become a scalable innovation, it must pass what we call the Big Bang Innovation test: it should improve existing solutions tenfold (10X or more), and provide lots of opportunities for the *creative crowd*. Otherwise, an inventor or entrepreneur is destined to fight an uphill battle against established industry players on the incumbents' turf, wasting his or her chances for success.
3. On the personal level, we see innovation-related creativity as a process of performing at the top of one's abilities and accomplishing *tangible results* on the path from invention to innovation, rather than experiencing a subjective, fleeting good feeling of *being creative*. An innovator can more effectively generate tangible results by focusing on developing the ability to match specific problem-solving skills to high-value problems. We devote the rest of the book to developing the reader's skills that improve the ability to identify and solve high-value problems/opportunities.

Based on our own invention/innovation practice and years of teaching experience, we don't expect the reader to learn everything at once. Different people have different approaches to learning. Some of the tools we offer in the book are analytical, while others are more concrete and focused on a given task. The best way to use the book is to read it entirely, find a subset of methods that best fit the reader's thinking style, and start applying them in real life. To facilitate the process, we provide case studies and review questions (some of them could be difficult to answer in the beginning). As readers improve their problem-solving techniques, they can revisit specific topics in the book.

From an innovator's perspective and in contrast with working on traditional school tests and puzzles, most of the problems one can solve are not worth solving. Finding just one *worthwhile* challenge and addressing it systematically, with the passion and the right skill set, can lead to a major breakthrough.

REFERENCES

1. Acemoglu, Daron, *Why Nations Fail: The Origins of Power, Prosperity and Poverty* (New York: Crown Publishers, 2012).
2. *The Wall Street Journal*, December 17, 2012. http://online.wsj.com/article/SB10001424127887324677204578185552846123468.html
3. Becket, Janie, "Study Shows Stanford Alumni Create Nearly $3 Trillion in Economic Impact Each Year," *Stanford News*, October 24, 2012, http://news.stanford.edu/news/2012/october/innovation-economic-impact-102412.html
4. Cowen, Tyler, *The Great Stagnation* (New York: Penguin, 2011).
5. Gordon, Robert J., Is U.S. Economic Growth Over? Faltering Innovation Controls the Six Headwinds. 2012. NBER Working Paper 18315.
6. Christensen, Clayton, "A Capitalist's Dilemma, Whoever Wins on Tuesday," *New York Times*, November 3, 2012, http://www.nytimes.com/2012/11/04/business/a-capitalists-dilemma-whoever-becomes-president.html?smid=pl-share
7. Foremski, Tom, "If Steve Jobs Were Starting Out Today He Would Struggle to Get Funding: He's a Marketeer Not an Engineer," *ZDNet*, November 3, 2011, http://www.zdnet.com/blog/foremski/if-steve-jobs-were-starting-out-today-he-would-struggle-to-get-funding-hes-a-marketeer-not-an-engineer/2010
8. West, Geoffrey, "The Surprising Math of Cities and Corporations," *TED*, July 2011, http://www.ted.com/talks/geoffrey_west_the_surprising_math_of_cities_and_corporations.html
9. Kahneman, Daniel, *Thinking Fast and Slow* (New York: Farrar, Straus, and Giroux, 2011).

Prologue: Unlearning What's Untrue

> The most useful piece of learning for the uses of life is to unlearn what is untrue.
>
> **—Antisthenes (445–365 BCE)**

BARRIERS TO CREATIVITY

In the world of innovation, we run into two kinds of barriers. Some of them we call hard, the other soft. Hard barriers are resource related. They feel like externally imposed limits on what can be done given the circumstances. By contrast, soft barriers are constraints of our minds. They are difficult to overcome because we do not see them as barriers in our own thinking. A lot of what we learn about invention and creativity through formal education and professional training builds soft barriers instead of removing them. In the Prologue, we will try to undo some of the damage.

HARD BARRIERS

Hard barriers to innovation are easy to discover. Here is a list of the most common ones:

1. Lack of physical resources
2. Absence of an essential implementation technology
3. No market for a novel product
4. Opportunity-poor environment

Hard Barrier 1: Lack of Physical Resources

Christopher Columbus (1451–1506) lacked physical ships and money to hire a crew and undertake a westbound voyage where he had hoped to discover the riches of Asia. After spending years lobbying private investors, governments, and kings and queens of Europe for ships and money, Columbus succeeded in convincing Ferdinand II, the king of Spain, to support the journey. When the sailor-entrepreneur threatened to take his project to the king of France, Spain's monarch relented and agreed to fund three quarters of the least expensive plan. Columbus seized the opportunity by providing (most likely, borrowing) the remaining quarter himself.

On August 3, 1492, his three small ships—the Niña, Pinta, and Santa Maria—sailed west, toward what he believed to be Asia. Just over two months later, on October 12, the expedition discovered what we now call the Bahamas, a group of islands in the Atlantic Ocean not far from the Florida peninsula. Although it took decades of follow-up expeditions to discover the rest of America, Columbus was the one who unknowingly and mistakenly started the process of globalization. If he did not get the resources to fund the expedition, he would fail to create new opportunities for trade and growth.

His success inspired others to seek fame and fortune in the New World. In 1606, Captain Christopher Newport, on behalf of the Virginia Company, led three ships with European settlers toward the west. After a successful voyage, they founded Jamestown, the first English colony in North America. Although they had adequate resources to make the trip, most colonists died within one year. Astonishingly, they starved among abundance. Rivers in the area were full of fish and local native Americans were friendly, ready to trade food for manufactured goods. But the settlers came to the new continent with a get-rich-quick mindset. Their approach—modeled after Spanish conquistadors' tactics— was to defeat the natives militarily, take their gold, and force them into slave labor. It didn't work. The local tribes were strong, and besides, they did not have the gold. Nevertheless, during the next several years the Virginia Company sent more ships and settlers to North America with the same bad idea. As a result, over 80 percent of colonists died from disease and starvation. What killed them was their faulty thinking (the soft barriers in their minds), not the lack of physical resources.

Hard Barrier 2: Absence of an Essential Implementation Technology

In the nineteenth century, English inventor Charles Babbage (1791–1871) came up with the idea of a mechanical computer, or as he called it the *Difference Engine*. He spent years trying to build the machine, but failed because manufacturing technologies at that time were not good enough to produce the precision parts he designed. Using the services of human computers (people who performed calculations manually) turned out to be more practical than building a complex, highly specialized mechanical device.

Babbage never fulfilled his technical vision, but in 1991, two hundred years after the inventor's birth, a working Difference Engine number 1 was produced and installed in the London Science Museum. During the years after Babbage's original invention, the manufacturing problems disappeared and so did the need for the large-scale mechanical calculators he invented. In the end, the idea of automated computing proposed by the inventor turned out to be prophetic. Unfortunately for Babbage, the technologies required for implementing the invention did not materialize during his lifetime.

Let us compare this situation to the story of negotiations between two young inventors and Excite, Inc., then a highly successful Internet search portal. In 1997, reluctant to start their own company, Larry Page and Sergey Brin, then Stanford University graduate students, offered Excite an exclusive license to their BackRub search engine, which used novel PageRank technology. The inventors wanted $1.6M, but Excite's tentative counteroffer amounted only to $750K and the proposed deal fell apart. In 1998, Page and Brin founded Google, Inc., which a few years later turned into a dominant force in the search business. In 2001, after the burst of the dot-com bubble, Excite filed for bankruptcy and the court sold the company's assets for $2.4M. To a large degree, a lack of creative thinking (a soft barrier) in their direct field of technical expertise, not the lack of a superior search technology, was the cause for Excite's demise [1, p. 30].

Hard Barrier 3: No Market for a Novel Product

In the beginning of the twentieth century, an entrepreneur and future member of the American Inventor Hall of Fame Clarence Birdseye (1886–1956) came up with the idea of commercial distribution of frozen foods. The invention occurred to him when he was watching Eskimos flash-freeze fish by throwing them on ice immediately after they caught them. The Eskimos' method preserved the taste and texture of the fish, compared with the more common process of drying or slowly freezing food. Nevertheless, despite the clear taste advantage of the new technique, frozen food had limited consumer appeal. The market for it emerged half a century later when the electric refrigerator (a complementary technology) became a common supermarket and household appliance.

Beginning in the 1960s, frozen foods attracted an even greater number of consumers due to the emergence of family units with two working parents and the invention of the microwave oven. The industry envisioned by Birdseye in the beginning of the twentieth century flourished several generations later when the market conditions caught up with the inventor's original vision.

Hard Barrier 4: Opportunity-Poor Environment

In *Outliers,* best-selling author Malcolm Gladwell documents remarkable stories of creative success and failure, including the one about Christopher Langan (born c. 1952), "the public face of genius in American life, a celebrity outlier" [2]. Langan has an intelligence quotient (IQ) of 195, which is 45 points or 30 percent, higher than that of Albert Einstein, who is widely considered the greatest intellectual of the twentieth century. In Gladwell's book, Christopher Langan competes on the *Who Wants to Be a Millionaire* show, winning handily a large sum of money in a One versus One Hundred contest. The episode looks like a triumph of a creative genius [2].

As the story progresses, we begin to realize that Langan's intellectual success is rather superficial. Due to unfortunate family circumstances and poor social skills, he never finished college (although his high school grades and the general level of his learning abilities were way above his peers). The lack of a college degree confined him to jobs that did not provide the intellectual challenge required for creativity development. Despite his extraordinary intelligence, Langan ended up working menial jobs: a bouncer at a wine bar, railroad worker, and the like. The incredibly high IQ failed to help Langan overcome the deficiencies of an opportunity-poor environment.

In addition to difficult personal circumstances, corrupt and inefficient government institutions are often responsible for a lack of opportunities. Economists Daren Acemoglu and James A. Robinson show that political and economic constraints stifle innovation and entrepreneurship. Comparing business environments in the United States and Mexico, they write:

> If you're a Mexican entrepreneur, entry barriers will play a crucial role at every stage of your career. These barriers include expensive licenses you have to obtain, red tape you have to cut through, politicians and incumbents who will stand in your way, and the difficulty of getting funding from a financial sector often in cahoots with the incumbents you're trying to compete against. [3]

Although frustrating, these and other hard barriers discussed previously are easy to discover because we bump into them head on. Later in the book, we will describe methods that help break them. In the meantime, let's turn our attention to soft barriers, the barriers that exist only because we think they exist.

Soft Barriers

Soft barriers are difficult to overcome because they lie below the threshold of one's perception. When we run into a hard barrier we know almost immediately that something is wrong. On the other hand, when we encounter a soft barrier we often feel nothing. The obstacle is perceived as a natural part of the world, and unless we spend a conscious effort examining our own thinking, we never discover what prevented us from making a breakthrough.

Often, soft barriers are perception based. For example, common negative stereotypes can induce a lack of confidence that harms performance on mental tasks. According to researchers,

> Even groups who typically enjoy advantaged social status can be made to experience stereotype threat. Specifically, White men perform more poorly on a math test when they are told that their performance will be compared with that of Asian men, and Whites perform more poorly than Blacks on a motor task when it is described to them as measuring their natural athletic ability. [4]

In other words, believing you cannot perform on a task makes you fail on the very same task. This psychological barrier is so difficult to overcome that Randolf M. Nesse, a professor of psychiatry and psychology at the University of Michigan, thinks that people who believe in things they cannot prove, have selective advantage over their peers [5]. This does not mean that to produce a successful innovation, we should become irrationally optimistic about our own abilities. On the contrary, research shows that people are overconfident on easy tasks, and underconfident on difficult tasks [6]. To succeed, we can and should use our rationality to systematically expose and overcome the soft barriers that prevent us from creating breakthroughs.

Three Soft Barriers to Overcome

The following three barriers are particularly harmful for inventive thinking. They become lodged in our minds during the years of exposure to standard education, popular culture, and mass media. Most of them are common misconceptions derived from limited or obsolete knowledge.

1. Everything is a trade-off and there's no such thing as a free lunch.
2. Invention is the same as innovation.
3. Creativity is a personal trait, not a skill that can be taught.

Soft Barrier 1: Everything Is a Trade-Off

Trade-Offs: From Kindergarten to MBA

For many of us, making trade-offs is a natural part of everyday experience. It starts in early childhood. "You won't get your dessert until you eat your dinner," one of the first rules children hear from adults.

Later in life, after getting our first allowances, we learn that the nicer the thing is that we want to buy, the more it is going to cost us. Over time, this seemingly immutable relationship between quality and price becomes so embedded in our thinking that we automatically associate a higher price with better quality.

In a psychological experiment, researchers at MIT provided one group of people with a painkiller and told them that it cost $2.50 per pill. Then they gave the same painkiller to a different group and told them that it had been discounted to $0.10 per pill. In both cases the researchers used the same placebo pills, but the people who took the "expensive" medicine felt a lot less pain than their peers in the "cheap" condition [7].

In another experiment, Dr. Antonio Rangel and his colleagues at CalTech gave people a taste of five Cabernet Sauvignon wines ranging in price from $5 to $90 a bottle. Personal interviews and brain scans confirmed that people appreciated the taste of expensive wines better than the cheap ones. The caveat was that the researchers used only three wines, serving two of them several times at different prices [8]. Somehow, paying more for a higher-quality product feels natural. Over time, we learn to accept and optimize options within a number of other common trade-offs: work versus leisure, risk versus reward, quantity versus quality, online convenience versus privacy, flexibility versus fuel efficiency (see Figure 0.1).

Trade-offs are so ubiquitous that "Everything is a trade-off" is taught as the first principle of economics. In the bestselling *Principles of Economics*, by Harvard professor N. Gregory Mankiw, right after a brief introduction into the discipline we read, "To get one thing that we like, we usually have to give up another thing that we like" [9].

As another example, *Policy and Choice*, a book by three highly regarded economists specializing in public finance, mentions trade-offs on 44 out of its 200 pages. Here's how it typically sounds:

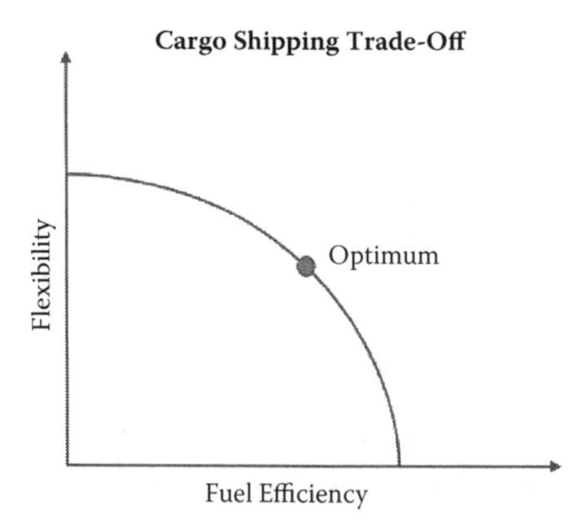

FIGURE 0.1 Shipment flexibility versus fuel efficiency trade-off diagram. Cargo shipping by sea is more fuel efficient than by airplane, train, or truck, but the shipment may take longer to arrive, and possible destinations may be limited to major ports.

> Many of the trade-offs that policymakers face are specific manifestations of a general, pervasive trade-off in public finance between equity and efficiency. For example, social insurance provides income-smoothing benefits but can induce costly moral hazard. [10]

When we talk to a well-educated policymaker or a business manager, we commonly find that the trade-off-based thinking is ingrained in his mind. By the time he completes an MBA degree or receives numerous promotions in the government and industry, he knows by heart that "there is no such thing as free lunch."

Engineering Trade-Offs

Making trade-offs while engineering a product is a common practice as well. In *Essential Engineer*, a book by Henry Petroski, the Aleksander S. Vesic Professor of Civil Engineering and professor of history at Duke University, we read, "Engineering is all about designing devices and systems that satisfy the constraints imposed by managers and regulators" [11].

Much of historical research into engineering solutions shows deep roots of this tradition. In 500 BCE, military engineers of ancient Greece faced a number of trade-offs while designing catapults for throwing large rocks. On the one hand, to destroy an enemy fortress they wanted to use the heaviest rocks they could find. On the other hand, a very heavy rock would not fly far and would force them to position the catapult close to the enemy. Due to the short range, the catapult's crew would be in constant danger of hostile arrows and small-rock fire. Because the lives of slave crews were cheap, the ancient Greeks decided to sacrifice them for the sake of the high-impact destructive power of the large rocks [12].

We find engineers facing similar trade-offs 2,500 years later. Consider state-of-the-art Internet data centers that comprise millions of processors collectively performing trillions of operations per second. In the process, they generate a lot of heat. The data centers require major cooling efforts to prevent the hardware from overheating. To address the problem, Google engineers have decided to house one of their new powerful data centers further north, in the building of a former Finnish paper mill factory. The location allows them to cool the equipment with seawater. Although the data center is located far away from most Google customers, the engineers seem to be willing to accept the trade-off.

Just like their management peers, engineers are trained to create and accept trade-offs. In *What Engineers Know and How They Know It*, Walter Vincenti, Professor Emeritus of Aeronautical and Aerospace Engineering at Stanford University and member of the National Academy of Engineering, teaches designers that during the engineering process, "Numerous difficult trade-offs may be required, calling for decisions on the basis of incomplete or uncertain knowledge" [13].

The younger generation of professors follows similar logic. A quick search over syllabi in premier American universities produces hundreds of hits for "engineering is about trade-offs." In a course for electrical engineering and computer science undergraduates at UC Berkeley, Dr. Dan Garcia emphatically brings to his audience's attention the inevitable trade-offs computer designers face all the time: performance of hardware versus flexibility of software, processing power versus energy consumption, low cost of disk storage versus high speed of random-access memory, and so on.

To illustrate the point during one of the lectures, he tells a clever story about a powerful king who asked his court scientists to distill into one sentence everything there is to know about the principles of good governance. After years and years of research (so the tale goes) the advisers come back and tell the king their conclusion, "Everything is a trade-off and there's no such thing as free lunch." Impressed with their laconic wisdom and powerful examples, the king embraces the principle, governs accordingly, and as a result, everybody in the kingdom lives happily ever after.

Most of us (educated through school, workplace, and everyday experience) are trained to appreciate and accept trade-offs. Furthermore, because managers, engineers, and government officials hold college degrees, we know they had to learn and internalize the "everything is a trade-off" doctrine. In a world like this, trade-off solutions have become a self-fulfilling prophecy. Managers ask for them, engineers deliver them, and customers accept them. Thus, the decisions of the mind made under limited or uncertain knowledge turn into physical constraints, forcing us to make future trade-off decisions. Such trade-offs happen because we believe they ought to happen.

Breaking, Instead of Making, Trade-Offs

If everybody is accepting trade-offs and premier educational institutions are teaching how to make them, is there a good reason that inventors should doubt this particular wisdom of crowds? Is there anything wrong with thinking that everything is a trade-off and there's no such thing as free lunch? Why should we consider it to be a barrier to innovation?

First, judging from our own personal experiences, the principle is not universal. For instance, we get a great "free lunch" when the best world universities—Stanford, MIT, UC Berkeley, Yale, Oxford to name just a few—provide us with their online courses for free. To partake in this educational feast, one only needs a low-end computer, an Internet connection, and a browser (or a mobile app). By contrast, studying the same academic materials in person would cost tens of thousands of dollars per year.*

Moreover, this opportunity offers another free lunch. Thanks to a software audio-video player built into our PCs and smartphones, we can listen to the online courses while doing chores, exercising, or walking the dog. That is, we not only have the ability to learn from the best of the best, but we can do it during the time that otherwise would be wasted. In short, no trade-offs are necessary.

Does this apply to this particular experience only, or is there a more general pattern here? To find out, let us briefly consider some breakthrough inventions that shape our everyday lives.

The Automobile Industry

In the very beginning of the twentieth century, people thought the automobile was an expensive toy for the wealthy. It was an unreliable, capricious technological curiosity. During the manufacturing process, highly trained mechanics assembled cars from individually made parts. To ensure that every part fitted properly, the mechanics had to make multiple adjustments. The more time they spent producing and assembling the parts, the more reliable the final product was. Thus, a common trade-off thinking at the time was, "the more reliable the car you want, the more money you are going to pay for it."

All that started to change in 1906, when Henry Ford introduced a breakthrough automobile design—the Model T. In combination with new mass manufacturing methods, the car broke a widely accepted trade-off. Ford cars were of highest reliability and much lower in price than any comparable model from his competitors. Ford automobiles became affordable to the American working class, something that could never happen within the framework of common thinking. By obliterating the widely accepted trade-off, the inventor initiated a revolution that reshaped the industry, the country, and ultimately the world.

We should also note that Ford imposed his own trade-off. He famously said that consumers "could have a car of any color, as long as it's black." This limitation was caused not by the inventor's particular love for black, but by the absence of color car enamels suitable for the new mass manufacturing process. As a result, to buy an inexpensive reliable car, one had to give up personal preferences.

* Or at the very least for the majority of standard courses one can do over 80% of learning for less than 0.2% of the cost of in-person attendance.

Unfortunately for the inventor, he insisted on the trade-off for too long. As the technology and market developed, so did consumer demands. In the 1920s, DuPont, a chemical company working together with General Motors, invented paints and processes that allowed GM to break through Ford's trade-off and take away a significant portion of the growing market. In a relatively short time, with additional technology and business innovations, GM overtook Ford as the largest car manufacturer for good.

In both cases, competitors leapfrogged one another when they broke through a trade-off created by the other party and accepted by the market as the natural state of affairs.

Shopping

Invention of the mass production technologies and related business methods was just one of the breakthroughs that shaped twentieth-century America. Let us consider another quintessential American experience—shopping.

Before the invention of the shopping cart, most people shopped with a hand basket. The more one bought, the heavier the basket became, and the less convenient it was to carry around. Also, if one wanted to buy a lot of goods at the store, the size of the basket would become a limiting factor. On the other hand, if the store owner introduced large baskets, carrying one around, with or without lots of purchases, would be an inconvenience. It appeared that changing any part of the system could only make things worse.

By the middle of the twentieth century, shopping was predominantly a women's affair. Upper-class ladies could afford servants who would help them with groceries, but working-class women had to do it themselves. Going to a local store every day was acceptable or even a pleasant social occasion for stay-at-home moms or single girls. But after America entered the Second World War and many men left home to join the army, women entered the workforce en masse, working long hard hours on jobs traditionally reserved for men [14]. Going to the store in the evening after work, carrying around a heavy shopping basket every day quickly became a tiring experience.

That's when a 1936 invention by Sylvan Nathan Goldman became quite useful on a large scale. Goldman was the owner of the Humpty Dumpty store chain in Oklahoma City. Some time that year, he noticed that the trade-off between the size or heaviness of the basket and the amount of goods the shopper could buy was limiting his potential revenues. The breakthrough came when he envisioned a large mobile basket carrier on wheels, which could hold lots of purchases and didn't burden the shopper with extra weight. Goldman's first shopping carts were basically rolling folding chairs with large baskets attached at the top (see Chapter 15, pp. 140–141). Introduced at

Humpty Dumpty stores in 1937 and patented by Goldman in 1940, the carts became a local business success.

As the United States pulled out of World War II and the postwar recession, the society became more affluent. In the 1950s, 1960s, and 1970s, people bought cars, built suburban homes, and shopped at supermarkets and stadium-size warehouses filled with a wide variety of goods. From a humble wooden contraption wheeled around in narrow aisles of a small grocery store, the shopping cart turned into a steel workhorse of the world's retail commerce worth trillions of dollars a year. None of the large-scale all-you-can-buy shopping experiences of today would be possible without Goldman's invention that broke the trade-off between shopper's convenience and the amount of goods one could buy at the store.

Internet Advertisement

On June 26, 2000, Yahoo and Google announced their partnership under which Google became the default search engine provider for Yahoo [15]. The next year, Yahoo paid Google $7.1 million for handling search queries from all Yahoo's web properties [16]. In 2001, in the aftermath of the spectacular burst of the dot-com bubble, Yahoo stood as a towering giant among Internet companies struggling for survival. The web portal was the largest source of search requests on the Internet. Because the company paid its search engine providers on a per-query basis, competition for the outsourcing contract was intense. Google won the business for the second year in a row due to the quality of its relevancy-based search algorithm, the approach Google founders Larry Page and Sergey Brin developed during their graduate research at Stanford University.

Impressed with the technology, Yahoo reworked its search results pages so that link rankings were essentially the same as Google's. The improved results drove up Yahoo search traffic, which annoyed Yahoo management. The reason was purely financial: more traffic meant higher payments to Google and less revenue for Yahoo. To reduce costs, Yahoo agreed to display the Google logo on its site and direct users to Google's own website. In other words, Yahoo directly helped to create a powerful competitor, who eventually replaced the company as a dominant force on the web.

Why did it happen? With search results being the same from both companies, the key difference was Yahoo's business model. As was its practice over the years, the company earned money by serving banner ads on its web properties: the more pages, the more links to those pages; the more banner ads per page, the higher the advertisement revenue. Of course, the space taken by the ads detracted from user experience (people hated the ads), but that was the price users were willing to pay to access free web content.

Fundamentally, Yahoo's business model contained an inherent trade-off between money made on ads and user satisfaction.

By contrast, Google earned its living by selling search services and providing relevant text ads through the AdWords service, launched in 2000 with 350 customers. First, ads served on Google pages along with search results were relevant and therefore useful to the users. Second, text ads didn't take much space on the page at all. Third, unlike Yahoo, Google's business model didn't provide any incentive for putting extraneous links on the page. Besides, the company was so frugal that it decided not to hire designers for their website. As a result, compared with Yahoo, Google pages looked Zen-like and were clutter-free.

Over time, it became a no-brainer for web users to choose between Yahoo and Google. The search results were practically the same, but the quality of experience was much better with Google. By breaking the trade-off between revenue and user satisfaction, Google destroyed Yahoo's business model and came to dominate the Internet advertisement business, reporting over $43 billion in advertising revenue in 2012 compared with the $7.1 million earned from running search queries for Yahoo during the whole of 2001. Taking the trade-off for granted eventually turned Yahoo from an Internet giant into a fumbling also-run.*

Accepting Trade-Offs Means Missing Out on Major Opportunities

As we can see from the examples, accepting trade-offs often means missing out not only on innovation, but also on investment opportunities. These kinds of opportunities may come about only once in a lifetime. Nevertheless, we are regularly taught in engineering and business courses that making trade-offs is the best we can do.

Had the innovators who created the technology and business breakthroughs discussed previously accepted the trade-offs widely considered as inevitable, they would have had a hard time succeeding on a large scale. The trade-offs stood in the way of extending their solutions to billions of people who faced difficult choices between quality and price, convenience and shopping needs, company revenue and user experience.

By breaking the trade-offs, the innovators made choosing new solutions a no-brainer. Of course, people wanted to have a quality product *and* a low price. Of course, people didn't want to limit what they bought at the store by the number of pieces they could physically carry. Of course, when they looked for information they didn't want to see annoying banner ads. Many trade-offs are well justified, especially in the short term. But in the long

* http://investor.google.com/financial

term they will be either broken or become irrelevant through the work of innovators.

Today, it is a widely accepted trade-off that the more environmentally sensible (greener) energy is, the more expensive it should be. For example, in 2007, the government of Spain guaranteed solar energy producers prices ten times higher than average wholesale price customers paid to mainstream energy suppliers. Over the three years that followed, the government "saddled Spain with at least 126 billion euros of obligations to renewable-energy investors" [17]. In 2010, in the aftermath of the financial crisis when it became practically impossible to borrow more money to support high levels of sovereign debt, the whole program collapsed, leaving many people—workers, entrepreneurs, bankers, manufacturers—broke. The initial belief in the trade-off that greener means more expensive led to unsound business and technology practices, which resulted in an economic failure and the public's deep disappointment with proposed environmental policies. Simply put, conventional trade-off thinking proved to be the worst way to approach a difficult problem.

Contrary to widely held notions and standardized education, not everything is a trade-off and there is such a thing as free lunch. Moreover, economist Joel Mokyr, the Robert H. Strotz Professor of Arts and Sciences at Northwestern University, who studies the impact of technology on society writes, "Technological progress has been one of the most potent forces in history in that it has provided society with what economists call a 'free lunch,' that is, an increase in output that is not commensurate with the increase in effort and cost necessary to bring it about." Further, he quotes the work of economists Kamien and Schwartz, "who regard technological change as a 'trick' that makes it possible, when asked 'which one' answer 'both'" [18].

Other economists also find that trade-offs are not necessarily inevitable in public policy decisions. For example, a National Bureau of Economic Research paper notes, "Investing in disadvantaged young children is a rare public policy with no equity–efficiency tradeoff. It reduces the inequality associated with the accident of birth and at the same time raises the productivity of society at large" [19].

As innovators, our goal is to avoid common misconceptions and look for free lunch ideas. Thinking of green energy, if sunshine is free and wind is free, why then should the energy we capture from them cost us more, not less? By treating a widely accepted limitation as a soft rather than a hard barrier, we can gain an advantage because everybody else thinks it is impossible to break the trade-off.

Review Questions

Next, we offer some exercises in identifying trade-offs and discuss inventions as means for breaking trade-offs. The objective of these exercises is to train the mind to use a more systematic and rigorous language for describing opportunities for invention (and later, innovation). To successfully answer these questions, it may be necessary to perform some research. Hint: Thankfully, modern Internet search engines enable direct searches for quotes in many books, as well as rapid access to background information.

Question 1: In the book *The Nature of Technology*, Brian Arthur mentions a key trade-off of twentieth-century aviation technology:

> In the 1920s, aircraft designers realized they could achieve more speed in the thinner air at high altitudes. But at these altitudes reciprocating engines, even when supercharged with air pumped up in pressure, had trouble drawing sufficient oxygen, and propellers had less "bite." [20]

Exercise: Restate the key trade-off more specifically, mapping it in two dimensions, if possible. What is the key performance characteristic that improves but then begins to decline as some variable is increased? Draw the performance trade-off curve on a two-dimensional plot and label the axes. What was the invention that broke the trade-off? How was the trade-off broken or circumvented? Discuss the influence this invention had on the evolution of aviation.

Question 2: During the Crusades (1095–1272), armies from European countries traveled thousands of miles to invade the Middle East. The Order of the Temple (or the Knights Templar) became a trusted party in charge of guarding large transfers of coins and precious objects that paid for "supplies, equipment, allies, ransoms and so on" [21, p. 153]. The larger the volume of the transactions, the heavier the load, the slower and more dangerous the journey would become; a greater number of armed guards would be required to protect the valuables from bandits, pirates, and hostile troops.

Exercise: Restate the specific trade-off that limited the size of the individual transaction, as well as the total number of transactions for the Templars. What financial invention(s) did the Knights Templar use to break the trade-off?

Question 3: In online Lecture 7 on technology entrepreneurship, Stanford professor Charles Eesley discusses various ways to improve ("hack") brainstorming. For instance, he says that you might not generate great ideas if you define the problem too broadly—for example, "How can we improve the world?"—because this type of problem statement doesn't allow the team to focus on any specific solvable problem. Professor Eesley also warns against defining the problem too narrowly, because you might miss important innovation opportunities. Finally, he advises to define problems just right—not too broad, not too narrow [22].

Exercise: Restate the trade-off involved in generating optimal problem definitions to improve brainstorming outcomes, making sure to use two-dimensional plots, as you did for the previous exercises. Note that in the first two exercises, it was relatively easy to place quantitative scales on the axes (e.g., air pressure or altitude, or transportation costs in contemporary currency). How would you quantify the scales in this instance, and how would you recognize that you've chosen an optimal breadth/specificity of the problem statement? How would you break the trade-off?

Note that answering these questions could be difficult, indeed, because typical discourse on the topic of brainstorming and idea quality remains qualitative in nature. This isn't to say that the total number of ideas generated during a brainstorming session can't be quantified, or that a panel of experts couldn't offer their opinion on the originality of each idea. Such a quantification/evaluation approach may work well for relatively simple problems (for which the technique of brainstorming was originally designed), but in real life even experts are notoriously poor judges of true breakthroughs. Examples include the failure of dozens of investors to back Fairchild Semiconductors, or Cisco, or J. K. Rowling's *Harry Potter*, or many other seminal undertakings. Recent advances in cognitive science show that issues and phenomena that had been discussed previously only qualitatively could be quantified and even engineered. Following in the footsteps of cognitive scientists, in Section II of this book, we offer specific tools and rigorous means of identifying and breaking the specific trade-offs involved in navigating the breadth/specificity of problem definitions.

Soft Barrier 2: Invention versus Innovation

During ordinary reading, people distinguish words by recognizing their graphical components and interpreting the visual information in the context of an article or a book [23]. Because the words *invention* and *innovation* look similar, and because they are often found within the same context, we tend to confuse the two terms or use them interchangeably.

Moreover, the idea of innovation has acquired such a positive cachet that there's a temptation to use the term as often and as broadly as possible. For example, while most dictionaries define invention as producing something for the first time, Michael Vance of the Creative Thinking Association of America writes, "Innovation is the creation of the new or the re-arranging of the old in a new way" [24, p. 109]. Although this statement sounds very encouraging to almost any reader, it's obvious that installing a new barbecue

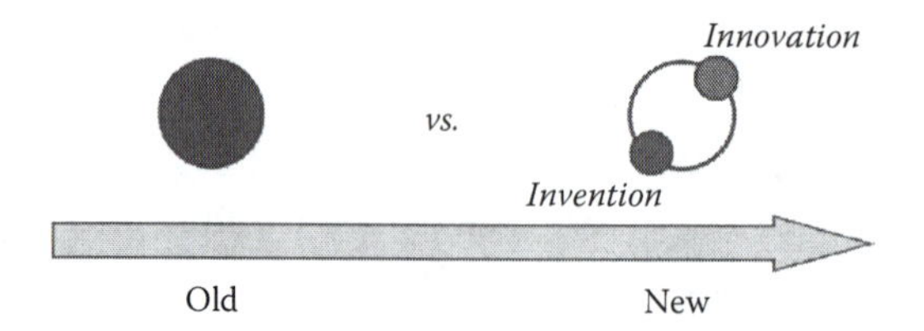

FIGURE 0.2 The general view of old versus new that conflates the concepts/meanings of the terms *invention* and *innovation.*

in the backyard or rearranging chairs on the deck is neither inventive nor innovative in any substantive way.

In popular culture, invention and innovation are often lumped together as an element of a bigger story about the battle between the old and the new [25]. Figure 0.2 illustrates this ten-thousand-foot perspective. The old is shown in black, the new is shown in white, with invention and innovation seen as small, barely distinct dots within the domain of the new. Such high-level contrast between the old and the new is useful as a writing device. It creates a sense of background tension that is essential for advancing the story. It also sidesteps the danger of overloading the general reader with a lot of technical detail.[*] But for inventors and entrepreneurs, understanding the details of how the new is brought into existence is essential for success. Therefore, we are going to zoom in on the new and try to understand the difference between invention and innovation, the two concepts with similar-sounding names, but with very distinct meanings.

Zooming In

Figure 0.3 shows a simplified two-dimensional map that can help us understand where we are and where we need to go in the white space of the new. The vertical dimension of the map ranges from *concept* to *it works* in terms of the functionality delivered to the user. The horizontal dimension ranges from *nobody* to *everybody* in terms of the ubiquity of the invention. The dot in the bottom left corner represents invention. It occurs when a person or a small group of people come up with an idea or an understanding of how a

[*] "In 1445 in the German city of Mainz, Johannes Gutenberg unveiled an *innovation* with profound consequences for subsequent economic history: a printing press based on movable type. Until then, books either had to be hand-copied by scribes, a very slow and laborious process, or they were block-printed with specific pieces of wood cut for printing each page. Books were few and far between, and very expensive. After Gutenberg's *invention,* things began to change. Books were printed and became more readily available. Without this *innovation,* mass literacy and education would have been impossible." (Acemoglu and Robinson, *Why Nations Fail,* 2012)

–A note to the reader: Acemoglu and Robinson use invention and innovation interchangeably.

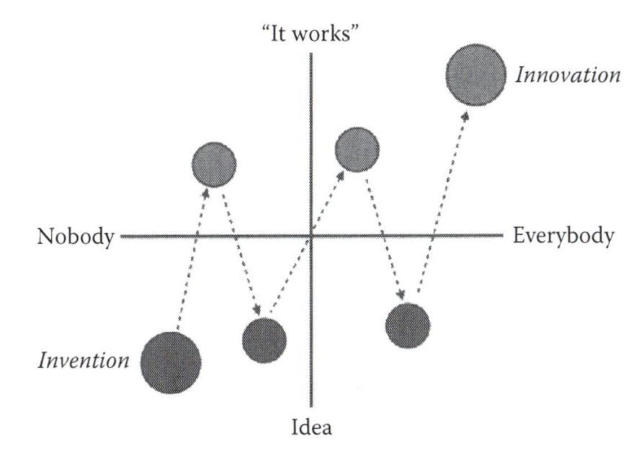

FIGURE 0.3 A more realistic, two-dimensional depiction of the difference (and path) between invention and innovation.

problem can be solved. Invention may also happen in the upper left corner when someone is tinkering with a gadget.

> Incredibly, Imo [the "monkey genius" of a colony of Japanese macaques] made two separate technical discoveries. First she discovered that to remove sand from potatoes thrown on the beach she could wash them in the sea rather than pick off the sand with her fingers. Then, in an even more remarkable display of ingenuity, Imo found that to separate rice from sand she did not have to pick out the individual grains; the mixture can be dropped into water where the sand will sink, and the rice will float and can be easily recovered. Both techniques were adopted by younger members of the troop as well as by older females and passed on to the next generation.
>
> —***Science and Technology in World History*** **[26, p. 8]**

To realize its full potential, the invention has to turn into something that works everywhere for everybody, that is, the dot in the upper right corner. For example, the Bardeen, Shockley, and Brattain 1948 invention of a junction transistor would correspond to a point in the bottom left corner. On the other hand, Intel's pursuit of squeezing billions and billions of transistors into a silicon chip and promoting the use of such chips in a wide variety of applications is a move toward the upper right corner, that is, the process we call *innovation*.

Invention is a personal or a small-group event. Innovation is a growing network of social events. One can be a great inventor, but a lousy innovator, and vice versa. The nature and the scope of problems encountered in the course of inventing and innovating are very different; therefore, we need to develop, learn, and apply different creative tools and strategies. To realize the potential of a breakthrough idea we, as inventors and innovators, have to be good at both activities. The path of innovation is not a straight line that takes you directly from point A (old) to point B (new) in the shortest amount of time. Instead, it involves a lot of zigzagging, multiple attempts at implementation, generation and validation of new concepts, reuse of existing technologies, and successes and failures.

Consider Johannes Gutenberg, the inventor of the modern printing press. He came up with his initial idea and built the first prototypes of the press in about AD 1440. Unfortunately, Gutenberg lacked the funds to develop the machine further, and in the period between 1450 and 1452 he borrowed 1,600 guilders from Johan Fust, a banker. By 1455, Gutenberg had failed to create a viable business. Fust sued and won a judgment for the repayment of 2,026 guilders: the original loan plus interest. In August 1457, seventeen years after Gutenberg's initial efforts, Fust and his partner published *Great Psalter*, the first commercially successful book, and the printing technology started its march along the path of innovation [21, p. 179]. Although the first block-printed books were produced in China before AD 800, that innovation never reached the impact of Gutenberg's idea. Gutenberg was a great inventor, but Fust managed to be a more effective innovator.

Today, more than half a millennium since Gutenberg's invention, the printing press is rapidly becoming obsolete. Paper books are losing their appeal, as electronic publications develop into a media of choice for the reading public. In Part III, we will explore e-book evolution scenarios in greater detail. In the meantime, we can predict with a high degree of confidence that the process of changing from printed to digital media will produce an impact of nearly infinite economic value, just as was the case with the innovation triggered by Gutenberg's invention.

Inventors and Innovators

New problems are discovered and new needs are created when ideas and implementations meet reality. Figure 0.3 shows this back-and-forth movement crossing the horizontal axis. In economics literature, this process is often called *diffusion*, a term introduced by Joseph Schumpeter, one of the first economists who studied innovation systematically. We don't like this term because it implies diffusion of the original idea embodied in an implementation, as if it emerged complete and whole out of the mind of its creator. In real life the process of innovation consists of many creative acts performed by many people at different stages of concept maturity. Schumpeter wrote in an age when the Gestalt theory of invention and heroic portrayal of an individual inventor was the norm. Since then, multiple studies have shown that the success of innovation critically depends on the contributions of numerous others—engineers, scientists, entrepreneurs, workers, bankers (yes, bankers also contribute to innovation), venture capitalists, lawyers, and so on. The existence of an environment rich in innovative contributors, be it before Industrial Revolution England of the eighteenth century, or postindustrial Silicon Valley of the twenty-first century, is absolutely necessary for a new technology to scale up. Any given company may play the roles of inventor and innovator during this process.

In our view, the process of innovation creates its own space. As an analogy, we will use the theory of an expanding universe. According to the theory and observations published in 1929 by Edwin Hubble, our universe does not stay in a permanent steady state as Isaac Newton and other classical physicists believed. Instead, the universe is expanding, with galaxies moving away from each other. The origin of the universe is an event physicists call the Big Bang. Does the universe expand (or diffuse) into a space that existed already? Not so, explains Stephen Hawking:

> The idea that the universe is expanding involves a bit of subtlety. For example, we don't mean the universe is expanding in the manner that, say, one might expand one's house, by knocking out a wall and positioning a new bathroom where once there stood a majestic oak. Rather than space extending itself, it is the distance between any two points within the universe that is growing. [27, p. 125]

In our case, Invention (Figure 0.3) is the Big Bang moment. From there on, the universe of innovation expands, creating new ideas, implementations, relationships, opportunities, and so on. In this perspective, innovations and technologies do not disrupt each other. Instead, they create new spaces for people and companies to move into and to develop further. Eventually, when a lot of people move into the new universe, the old one withers away or collapses.

For example, social networking created a space where people share digital pictures. Instead of accumulating prints in photo albums (heavy binders holding photographs printed on paper), users now post images online. As a result, the traditional photo industry, lead by Kodak, went bankrupt. In our view, it would be a mistake to describe the transition as a *disruption*, not only because nobody actively sought to destroy Kodak, but also because it limits our perspective to that of the old industry. A better idea would be to think of social networking or any other technology as a new space, where users can do a lot more than just share pictures. As inventors and innovators, we would be better off focusing on the development of novel opportunities for those who decide to move into the new space with us. The old space is worth further consideration only when we want to build user migration paths. Sometimes, such paths are not even necessary because a new generation of users may come to us directly, rather than from the "disrupted" world.

Figure 0.4 maps a company entry point into a particular innovation space. In 2011, Google, after multiple unsuccessful invention and innovation attempts with Orkut, Buzz, and Wave, copied Facebook's design and entered social networking space with Google+. A study from EyeTrackShop, a Swedish online user behavior research firm, showed that on Facebook and Google+ pages the way in which users scan the page was nearly identical [28]. On the innovation chart, Google's position is way to the right of the invention point. That is, not only is the concept of social networking not new, but it has been implemented previously by Facebook on a massive scale, with hundreds of millions of active users around the world. It should be clear that in the realm of social networking, Google's strategy so far isn't invention-centric, but is more biased toward leveraging its scale of operation and market power to result in innovation. The company has deep pockets and can afford subsidizing Google+ with revenue from its search-supported advertisement business.

The same can be said about Microsoft when it entered the advanced graphical user interface (GUI) PC market with the Windows 95 operating system. By that time, the process of innovation was well under way, propelled in the commercial space by Apple's Macintosh machines used by millions of consumers in educational and publishing markets. In both cases, Microsoft with

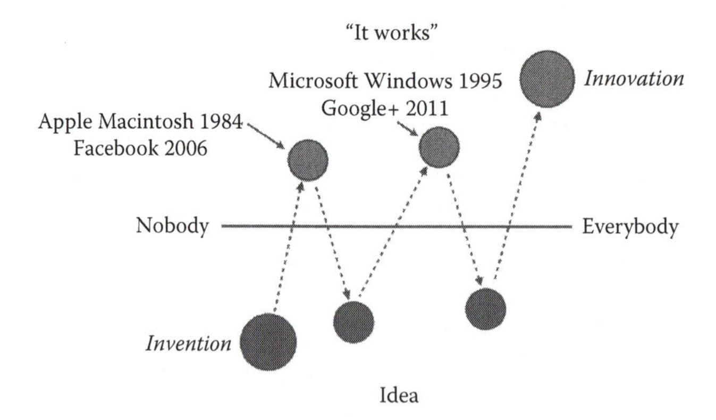

FIGURE 0.4 A company entry point into an innovation space. Apple and Facebook entered their respective innovation spaces closer to the invention point than Microsoft and Google+.

its Windows operating system and Google with its Google+ social network, aimed at expanding the appeal of the original technology and, if possible, taking prior users away from the existing implementations.

In contrast, Apple's entry into the GUI PC space happened much closer to the invention point. At the time Steve Jobs debuted the first Macs, the market for such GUI PCs simply did not exist. The GUI computer invented and built at Xerox PARC prior to the formation of Apple Computer Inc. was little more than a laboratory demonstration, with zero consumer market traction. The Xerox Star Information System, announced in 1981 with an estimated price of $75,000, was over ten times more than even the most expensive Apple machine ever sold to consumers [29]. Apple's explicit goal was to create a consumer PC market, with additional hardware, software, distribution channels, and other elements. The same applies to the iPod and iPhone. In contrast, Google's entry into the mobile space with Android was biased toward an innovation-heavy strategy. (Note again that the conflation of terms—innovation vs. invention—prevents us from understanding the true nature of the process. It would be fair to say that, e.g., Apple mixed in a greater amount of invention and innovation than Google in the mobile space. We understand this distinction as a 2-dimensional concept, rather than a simple 1-dimensional difference between "innovative" vs. "non-innovative." Recognize also that the business and technology risks in these two types of market entry, one on the left side and another on the right of the chart, are quite different, although the detailed discussion of this notion exceeds the scope of this book.)

If we use the same chart to represent social networking, lawsuits alleging Mr. Zuckerberg stealing somebody else's idea notwithstanding, Facebook entered the market in a manner more like that of Apple in 1984 than Google in 2010. If Facebook were diligent in filing for patents from the very beginning, Google+ would be in the same legal trouble as Google's Android customers are with regard to Apple's patent suits targeting mobile phone manufacturers.

A key difference between invention and innovation is involvement of various contributor communities. By contributing to the development of the original idea, they share the risks and rewards for being a part of an innovation wave. This broad-based buy-in accelerates and strengthens the pace of innovation (also known as the *bandwagon effect*).

The Creative Crowd

Figure 0.5 (a and b) shows a simplified diagram depicting the involvement of different types of contributors during the process of innovation. Similar to the previous chart, we have the horizontal dimension ranging from *Individual*

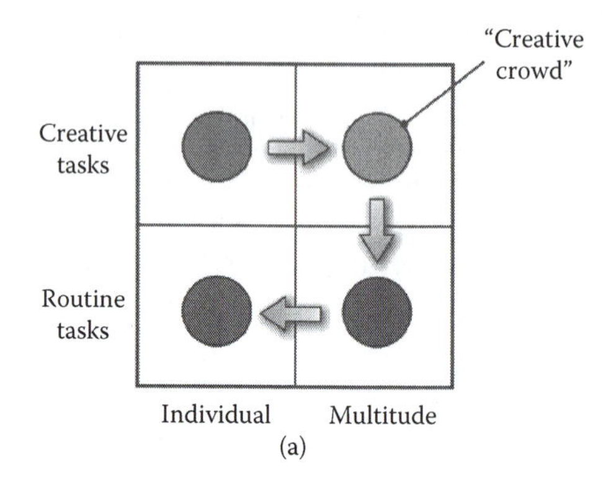

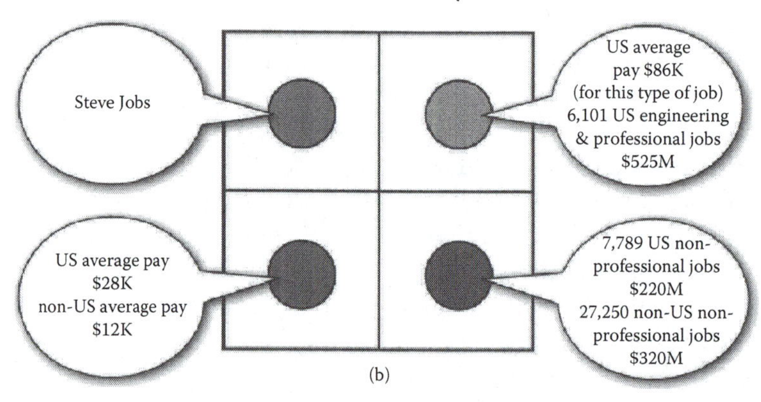

FIGURE 0.5 (a, b) Diagrams illustrating in graphical form the main conclusions of a recent study by Greg Linden et al., *Innovation and Job Creation in a Global Economy. The Case of Apple iPod,* 2011, http://pcic.merage.uci.edu/papers/2009/InnovationAndJobCreation.pdf. To earn good money, one must become involved in nonroutine processes in generating new products or services. Routine jobs pay little and are the target of relentless outsourcing, off-shoring, and, ultimately, automation. (For a stark illustration, see a related article by *The Week* editorial staff, "A Day in the Life of a Warehouse Wage Slave," http://theweek.com/article/index/228096/a-day-in-the-life-of-a-warehouse-wage-slave and the recent purchase by Amazon of a warehouse robotics system.)

to *Everybody.* The vertical dimension reflects the novelty aspect of the work people do, either in generating the concepts or implementing them. At the bottom of the vertical axis is *Routine* and at the top is *Creative.*

Here we need to clarify the terms used while examining the dynamics of the chart in Figure 0.5. According to the *Merriam–Webster Dictionary,* the word *routine* means a habitual performance of an established procedure that produces an expected outcome. Most innovations end up as routine procedures.

Using a shopping cart at a supermarket in the 1940s and 1950s was a novel experience, but now it is something performed habitually without a second thought. Similarly, using a virtual shopping cart on a website was a major novelty in the early days of Internet commerce, but now it has become routine as well. The same applies to browsing the web, sending an email, driving a car, using an ATM machine, and many other activities, once considered to be at the leading edge of innovation. Today, we take for granted technologies that in fairly recent human history were thought to be completely impossible. For example, the tragic deaths of Romeo and Juliet shown in a famous seventeenth-century play by William Shakespeare feel totally unwarranted to modern preteens who might ask, "Why didn't he tweet or text her before killing himself?" Today, we reasonably expect to have reliable means of communication available practically everywhere, including cities, rural communities, and even remote mountain paths. Eventually, successful innovation reaches everybody and becomes routine in its applications.

On the other hand, the word *creative* is associated with novel actions with uncertain outcomes. It implies trying (*trying* being the key word here) to accomplish something that is either totally new to the actor personally, or to everybody in the world. Creative actions carry the risk of total failure. Some people are more willing than others to take the risk and doggedly pursue a new concept or implementation. (We are not merely talking about *early adopters*—we are talking about early contributors who are taking on significant risks investing their time, money, and effort.) We will discuss creativity later in the book, but in the meantime, we will use this simple distinction between the notions of creative and routine.

Returning to the chart in Figure 0.5a, we begin in the upper left corner—Creative-Individual—at the moment when the individual or a small group of people develops a novel concept or a prototype. For example, it is often said that Henry Ford came up with the idea of a car assembly line when he observed a moving cattle processing line at famous Chicago slaughterhouses.* We don't know whether the idea came to him in a *flash of genius* or as a result of deliberate thinking, but we do know for a fact that it took him many years to develop a practical implementation. Assembling cars on a massive scale is a process that is radically different from slicing animals for meat, bones, and hides. That is, a new production process had to be invented and tested, a new car suitable for such a process had to be invented and implemented; a new kind of manufacturing equipment had to be built; new workers had to be trained; and many more details that make the original idea successful had to be created in real life. To accomplish all

---

* Rob Wagner, "Early Ford Model T Information," eHow, http://www.ehow.com/info_8516174_early-ford-model-information.html

of these creative tasks, each one of them rife with problems and uncertain outcomes, Henry Ford hired a group of talented engineers, managers, and mechanics to push his revolutionary enterprise along the axis of innovation. It took years before the slaughterhouse idea turned into an automobile factory. On the diagram in Figure 0.5a, we show this transition as a horizontal move from the upper left to upper right, into the Creative-Multitude quadrant, where the most difficult and exciting part of innovation actually happens.

The Opportunity

> Shakespeare, for example, worked at the center of a group of quite exceptional London dramatists. At different times, members of this group collaborated. They also all stole willfully from one another. Simultaneously, they all competed, driving each other to greater and greater heights. None of this limits Shakespeare's towering genius.
>
> **—J. S. Brown and P. Duguid, "The Silicon Valley Edge" [30]**

In their book *The Social Life of Information* John S. Brown and Paul Duguid wrote, "the challenge in moving systematically from an initial invention to innovation … is the challenge of coordinating diverse, disparate, and often diverging, but ultimately complementary, communities of practice" [31]. Who are these communities that take a roller-coaster ride of the initial stages?

Some of them have strong connections to the company that works on the original idea. For Henry Ford it was a small group of engineers, managers, and investors who poured their money and labor into the enterprise. One of them, Clarence W. Avery, guided the work on developing the assembly line and increasing its performance from 12.5 to 1.5 hours per car. This work was absolutely essential to the affordability of Ford's bestseller Model T. The Dodge brothers, John and Horace, were initial investors and suppliers of engines; John, the elder brother, served as a vice president in the Ford Motor Company. In 1914, to attract the best mechanics, Henry Ford offered to pay $5 per hour, *double* the prevalent wage at the time. All together, the main incentive for Ford's creative crowd was the opportunity to benefit from a newly created market for affordable automobiles.

Today, we can see introduction of a new product specifically designed to attract creative people and companies capable of adding value to the original concept. In 2007, when Steve Jobs introduced the iPhone, he made sure companies as diverse as Google, AT&T, and Disney provided an environment where the device could be successful, with software, connectivity service, and media content available to the users. Apple offered third-party software developers and content providers an opportunity to benefit from the popularity of the iPhone. Using iTunes and the App Store, the creative crowd could sell applications, music, and videos to the new generation of consumers. In other examples, companies such as Amazon and Facebook became incredibly successful by providing opportunities for creation and sharing of commercial and user-generated content. Some of those opportunities were economic, others were social, and yet others were revolutionary on the scale of an entire society.

Being aware of the difference between invention and innovation helps us develop inventions that either generate opportunities for others, or allow us to take advantage of a higher-level innovation *trend*, becoming an early member of the creative crowd. Eventually, the work of the creative crowd results in the development of largely risk-free products, services, and processes related to the original concept.

From Being Creative to Being Stuck

The move from the upper right quadrant to the bottom right—Routine–Multitude—means making the innovation part of a routine for lots and lots of people. Nearly always this is accomplished by building once-revolutionary ideas right into the infrastructure. For example, producing or driving a car used to be a challenge, but now it is a well-understood practice with firmly established rules and highly predictable outcomes. In the late 1980s, email was a novelty; now, it feels natural or even obsolete. Today, in the United States if you don't have access to a car or don't know how to use email, you are likely to have difficulties finding or keeping a job.

The transition from the lower right to the lower left in Figure 0.5 leads to situations where individuals feel like they have no choice but to accept the inevitable participation in certain technological processes. Operating within the two bottom quadrants in your professional life means low pay and very few, if any, creative opportunities [32]. In a developed society, not being creative often means being stuck within processes designed and defined by others. It entails a life within a rigid set of constraints, a multitude of tradeoffs, and doing what you are told to do. This existence appears to be practically risk-free until the predefined processes are made obsolete by external forces, either through creative destruction of business niches, outsourcing, or a recession. In such instances, people find themselves without jobs and in possession of suddenly obsolete skills.

Innovation is not neutral. Although society as a whole benefits from it, the process of innovation produces winners and losers, depending on where the person happens to be relative to the creative–routine divide. Paradoxically, being an inventor for an industry or technology domain that lacks opportunities for the creative crowd may not be a good idea. To be a successful inventor, one needs to have a critical mass of innovators around him or her.

Summary

Invention is the process of coming up with a new, useful, and nonobvious concept, either an idea or a prototype. *Innovation* is the process of making the concept work for a multitude of people in a wide variety of contexts. Invention is a personal or small-group process or event. Innovation is a network of social processes and events.

The challenges and problems one encounters during invention are different from those encountered during innovation; therefore, effective problem-solving methods may need to be different. Depending on where (along the zigzagging path from the small-scale extreme of invention to the large-scale innovation process) one's business enters the market, he must develop and deploy different strategies to gain a competitive advantage.

To succeed as an inventor or an innovator, either individually or as a company, one needs to provide opportunities for many other inventors and innovators, a social group of people we call the creative crowd. One can also succeed by becoming a member of the creative crowd, provided one works in a high-opportunity environment.

Review Questions

1. Draw a two-dimensional diagram that distinguishes between invention and innovation, and place points on it corresponding to the evolution of the printing press. What are the trade-offs and constraints that were being broken or circumvented with each invention event?
2. In your opinion, do the following sentences relate to invention, innovation, or both? Explain.
   a) "Before the invention of agriculture, food was obtained by fishing, hunting, and gathering from uncultivated trees, bushes, and plants" [33].
   b) "In 1445 in the German city of Mainz, Johannes Gutenberg unveiled an innovation with profound consequences for subsequent economic history: a printing press based on movable type" [3].
   c) I don't see any reason to reinvent the wheel.
   d) Nikola Tesla was a better inventor than Thomas Edison.

e) "Patent indicators are used to map aspects of the innovative performance and technological progress of countries, regions or certain specific domains and technology fields" [34].

f) "Leonardo Da Vinci—one of the greatest ever minds, invented models that proved workable 3–500 years later" [35].

Exercises

1. Let's establish the "Big Bang test" – does the innovation you are pondering pass the Big Bang criterion of creating a new universe, a new space-time?
2. In the film "Avatar" a different civilization is described. What would compel people in our current world to migrate over to the new world? What would compel people in that world to migrate back into our current world? Create two columns that would describe the Pros and Cons of migrating into one world, and the same for migrating into the other.
3. Often, inventors use old terminology to "trick" people into moving into the new world. An example of that is the term iPhone that Steve Jobs used to suggest that the device was a new kind of (smart)phone. But it was terrible at being an actual phone, while being much better at other things. Nowadays, users clearly spend far more time doing things on their iPhones other than traditional phone conversations. Can you think of other examples when inventors or innovators have to use old terminology to entice users into the new world?

Soft Barrier 3: The Myth of Creativity as an Innate Trait

Leonardo da Vinci versus Giovanni Battista Morgagni, a Martian's Perspective

Among the creative geniuses of the past, Leonardo da Vinci (1452–1519) is considered to be one of the greatest, if not the greatest, of all time. Compared to modern standards, the scope of his work appears enormous, earning him the title of the ultimate *Renaissance man*. He was a painter, engineer, inventor, architect, sculptor, writer, scientist, anatomist, and much more. Leonardo's paintings are the jewels of the world's best art collections. His portrait of Mona Lisa displayed behind thick security glass in the Louvre in Paris is not only the destination of pilgrimages among international tourists, but also the subject of many scientific papers trying to understand the mystery of the woman's smile. Leonardo's personal notebooks (written in mirrored script) covered a wide range of topics, including fantastic mechanisms: flying machines, steam cannons, submarines, tanks, and many others. To achieve the lifelike quality of his paintings, Leonardo studied laws of perspective, measured objects, dissected cadavers, creating in his notebooks first-of-a-kind anatomical depictions of human bodies and internal organs. For example, he was the first one to make scientific drawings of the heart, a fetus in utero, the wormlike appendix, and so on.

Leonardo's life was full of mysteries and unanswered questions. A man of very humble origins, he rose to keep company with kings and popes, witnessing and participating in many historic events, accounts of which he kept to himself. He wrote his notebooks in code and stashed them in secure places to thwart uninvited curiosity. Surrounded by followers and rivals, he managed to keep his life as private as possible, including his relationships with women and men. As a result, these relationships became a subject of rampant speculation lasting from the sixteenth to the twenty-first century. After his death and publication of the deciphered notebooks, Leonardo's art and personality created almost an industry of people setting out to uncover and understand the secrets of his genius. Today, there is no lack of authors, directors, and advertisers mining Leonardo's fame for entertainment, creating mystery novels, movies, and even sports shoe ads, featuring a bearded sage slam-dunking a ball on the basketball court. The name of Leonardo has become a symbol, a synonym for creativity and invention.

A. B. Usher, among many other authors, considered him the epitome of an inventor. Leonardo's ability to visualize and draw his insights provided psychologists and historians of invention with a powerful confirmation for the Gestalt theory: creative solutions appear to us as complete visual patterns. Insight is a work of genius, and even the staunchest proponents of the difference between invention and innovation acknowledge, "There is no known way to teach someone how to be a genius" [36].

Before we agree to the assertion above, let us consider the life of another Italian, Giovanni Battista Morgagni (1682–1771), a doctor, a teacher, and a genius. Unless you studied the history of medicine, neither his name nor what he did for humanity is known to you, even though the impact of his work is measured in lives saved—including yours and mine—reaching billions.

For at least two thousand years before Morgagni's work, doctors, philosophers, and laypeople alike believed that the cause of any human disease was an imbalance of four humors: black bile, yellow bile, phlegm, and blood. Accordingly, the treatment of a disease, even a deadly one like the plague, had to be directed at restoring the balance by purges, bloodletting, drinking beneficial substances, and by any other humor-restoring means. In the seventeenth century, Jean-Baptiste Molière (1622–1673), a French playwright satirized this "time-tested" practice in his 1665 play *L'Amour Médecin* (Love's the Best Doctor). In the play, famous doctors argue passionately whether the best way to cure a young lady of love would be bleeding her as much as possible or causing her to vomit extensively [37]. In short, the state of medicine when Morgagni joined the profession was a disaster.*

He was twenty-three, working at Bologna University as an assistant to Antonio Valsalva, a famous professor of anatomy, when the body of a man of seventy-two was brought in for an autopsy. The man, a laborer from Forli, Morgagni's native town, died after two and a half weeks of suffering from pain in the belly, fever, vomiting, and further pain that migrated from the abdomen down to his right leg, causing him to limp and rendering him unable to work. During the autopsy, Morgagni was taken aback by the horrible odor coming from inside the dead man's body. Cutting farther through the tissues revealed the shocking cause of death: a tiny wormlike structure in the intestines burst open, creating a massive inflammation that eventually consumed the man's life. It became crystal clear to the young doctor that no amount of liquid poured into or purged from that body could cure the disease. The man died of an organ failure, not an imbalance of humors [38].

The case described above was Morgagni's first encounter with the appendix. Two hundred years earlier, Leonardo da Vinci discovered and described this intestinal structure in his notebooks. He also noted that in some bodies the organ was normal, in some inflamed. But Leonardo's records were hidden and locked, and there was no way for Morgagni to read them. Morgagni consulted with his professor, and together with Valsalva they decided to autopsy all bodies coming to their hospital. In addition to that, they decided to keep detailed records from each autopsy along with patient case histories, which was a critical step that allowed doctors to correlate symptoms of a disease with the anatomy of organ failures revealed during postmortem examinations.

Fifty-six years later, when Morgagni was seventy-nine, he published five books with the results of this painstaking work, including hundreds of cases, with symptoms of many diseases indexed to their outcomes. The publication signaled the end of humors, and the beginning of new medicine, where symptoms were understood as "the cry of the suffering organs," organs to be treated by an attentive doctor with the proper knowledge of applicable therapies. Today, when you undergo your regular physical exam or even when you search online for symptoms for an ailment that bothers you, you reap the benefits of the medical revolution started by Giovanni Battista Morgagni, a genius of science, the inventor of an index that helps us hear and heed the cries of the organs.

---

* More than 200 years after George Washington's death, there is still a controversy whether he died from extensive bloodletting ordered by his doctor. http://www.newyorktimes.com/1999/12/14/health/death-of-a-president-a-200-year-old-malpractice-debate.html

A Martian's Perspective

Unlike da Vinci, Morgagni's personal life is no mystery at all—maybe that is why we do not hear much about his breakthrough work. A deeply religious man, he married early, and had with his wife, fifteen children—three sons and twelve daughters. His professional life was rather predictable too. He performed his autopsies, treated patients, wrote papers and letters consulting other doctors, taught students coming to his lectures from all parts of Europe, and received multiple international awards for his practical achievements. Even though his book, *De Sedibus et causis morborum per anatomem indagatis*, changed the course of medicine, saving an incredible number of lives along the way, you will not hear about this achievement as a major creative work. In contrast, the vast majority of Leonardo's technical concepts were never turned into practical devices, yet he is considered to be the epitome of a creative person. Why?

To answer the question, let's conduct a thought experiment. Imagine you are a Martian who considers the impact of inventions independently from the person or persons who made the invention. All humans look alike to him, but all inventions are unique.

First, you, the Martian, can see that humanity as a whole prefers inventions that generate high entertainment value to those that result in objective long-lasting positive outcomes. Accordingly, humans label as *creative* individuals with the highest amusement quotient, and regard creativity as something that, essentially, entertains the most. One Leonardo da Vinci produced wonderful paintings and sculptures along with sketches of the most amusing nonworking mechanical contraptions of all time. Therefore, he is considered to be more creative than one Giovanni Battista Morgagni, who developed a new approach to medicine. Selecting role models according to their entertainment value is great for teaching creativity to artists. This type of creative—in the artistic or aesthetic sense—work produces one-of-a-kind artifacts worthy of museums, books of fiction, and cinema. They stimulate imagination, inspire, and can be used for breaking psychological inertia.*

Nevertheless, the other type of creative—in the sense of shaping what hadn't existed before—work, where the inventor purposefully, but unassumingly, makes a scalable difference in the world remains obscured by the way in which humans tend to be ruled by emotions and quick intuitions in their daily lives (largely as a consequence of the relationship between the mental

* To be sure, works of artistic genius can inspire a great of deal of limitation and modification, spawn schools of particular kinds of painting, poetry, music, etc. Such "inventions" can be said to become aesthetic innovations.

System 1 and System 2, as described by Daniel Kahneman [39][*]). Creativity is presented in a way that appears to assume that how you feel about the person and his work is more valuable than the practical effect of this work on other people. *Interestingness* is elevated above usefulness and scalability, because often the latter is mundane, and it takes a conscious effort to discern the amazing in the mundane. A physical exam feels boring. Humans would probably appreciate Dr. Morgagni's work better if he made his discoveries by fighting zombies.

Second, you, the Martian, can see that even when humans celebrate inventors who achieved a long-lasting impact, they still focus on the amusing rather than the inventive aspects of their accomplishments.

Consider the case of one Thomas Alva Edison. Eighty years after his death, human children are still being taught that he invented the light bulb and said, "Genius is 1 percent inspiration, and 99 percent perspiration."

We humans seem to focus our attention on the light bulb, because it lights up and is hence easy to see. "Why did Edison need to invent the light bulb?" we ask rhetorically. "Because without the light bulb, you cannot have electric light and it would be dark. For Edison, inventing the light bulb was 1 percent inspiration, and the 99 percent perspiration went into trial and error of finding the right filament for the bulb," we say.

Since you, the Martian, can see things at a larger scale, both in time and space, and your vision is not affected by the electric light, you know that light bulbs existed before Thomas Edison. Moreover, a few years prior to "inventing" the light bulb, the same Thomas Edison used one to bring electric lights into the house of his friend and supporter J. P. Morgan. Confined in our thinking to the boxes of our homes, most humans do not realize that the challenge Thomas Edison faced was on a scale much larger than the light bulb. That is, the inventor had to bring lights to Manhattan, a big section of a large city, not just one place, for which existing light bulbs would be a sufficient solution.

The new big task required a big investment, with the largest chunk of it being the cost of copper wires necessary for delivering electricity from generators to thousands of light bulbs in the city. Before Edison, engineers thought that to feed that many lights, one needed high electric current running over thick wires. Edison, however, came up with an elegant and simple

[*] "I describe mental life by the metaphor of two agents called System 1 and System 2, which respectively produce fast and slow thinking. System 1 operates automatically and quickly, with little or no effort and no sense of voluntary control. The operations of System 2 are often associated with the subjective experience of agency, choice, and concentration." (Kahneman, *Thinking Fast and Slow*, pp. 13, 21, 2011)

wiring scheme that allowed him to use low current and thin wires, beating the copper cost constraint. That was Edison's 1 percent inspiration, an idea so brilliant that it prompted Sir William Thomson, one of the world's best physicists of the time, to remark, "No one else is Edison." After the breakthrough, Edison needed a light bulb with a different filament and it took him 99 percent perspiration over two years to find it. The inventor's creative logic is simple enough, but fooled by the propensity of our vision to focus on shiny things, most of us humans pay attention to light bulb stories, failing to recognize the significance of true breakthrough inventions, many of which have to do with infrastructure that remains largely invisible to us.

For example, invention of the transformer, not the light bulb, was the next major breakthrough because it allowed creation of large-scale electric networks that shaped American industrial revolution. In another example, Edison's phonograph, heavily promoted by the inventor and the media, was a curious device indeed. It could even record and play back "Mary had a little lamb." Unfortunately, it didn't work well enough to be used on a large scale. The real audio revolution started with the invention and subsequent improvements of the gramophone—a play-only audio machine with flat, sturdy record disks. Emile Berliner, the original inventor of the device, and Eldridge R. Johnson, an engineer who made numerous improvements to it, created a new way to share music that dominated the twentieth century.

With most of us so attracted to the personalities of inventors and stories about their lives, it is obvious to the Martian that we often lose sight of an invention's true creative impact. It should be obvious by now, that we do not know how to teach genius because we don't pay attention to what genius does.

Third, compared to creativity, the method humans use to teach important subjects is very different in effort and scope. For example, math is introduced early in elementary school and continued through at least a year or two in college. During the multiyear study, the focus is on practical work and its outcomes, not curious episodes and speculations about personal lives of numbers and theorems.* It is the challenges mathematicians encountered and solutions they developed, as well as applications of abstract theories to human practice, that help students better learn and use what they've learned in real life. Similarly, language arts are taught systematically for many years as a gradual practical experience with the magic of words, sentences, and literary works. Both in math and language, after many years of studies, students are not expected to become geniuses. Rather, they are required to

* This could be an interesting exercise in imagination development. See, for example, the famous book *Flatland: A Romance of Many Dimensions*, written by an English schoolmaster Edwin Abbot in 1884.

master certain essential usage skills and recognize incoherent math or language expressions, from functional ones.

In contrast, the creativity of invention is presented in an ad hoc manner. The students' attention is directed toward entertainment and standard courses propagate common sense confusions about trade-offs, teach that problems in need of solutions are merely puzzles, mix up invention and innovation, and so on. If somebody learns creativity through this process, it would be despite, not because of, education. In this environment, creativity cannot be learned because humans are being taught to move away from it. To change the situation, we need to make a vigorous effort to undo the damage.

The Science of Creativity: A Very Brief Overview

Richard E. Nisbett, psychologist, and distinguished professor at the University of Michigan–Ann Arbor, wrote jokingly about the general attitude toward creativity:

> It is hard to believe, but humans actually think there is a property of creativity that one can either "have" or "not have," as opposed to a talent for some field—a love of its content that keeps them thinking about it all the time—organization, and a willingness to work. And, of course, worries about his creativity will have the same salubrious effect on his work output that worries about his potency will have on his love life. [40, p. 24]

In a brief overview, we will try to capture some of the interesting findings that current science tells us about human creativity.

Research in the creativity domain is diverse, ranging from neuroscience and brain imaging, to analysis of personality traits, to large-scale population studies touching upon IQ and general intelligence factors. Perhaps one of the biggest challenges limiting progress in this field of study is the vague state of understanding of what creativity is. Though researchers commonly agree that creativity is one's ability to produce novel and useful results that are appropriate in a certain social context, there is a wide variety of interpretations of what each of the terms means precisely and how it can be measured [41]. For example, during brain imaging studies researchers present people with puzzles to see which brain regions become more active when the subjects are trying to solve them [42]. Although some of the results may bring relevant insights, arguably, such tests are far removed from what we consider a real-life problem-solving activity.

Our ability to become aware of a solution seems to be constrained by short-term memory, which can hold and manipulate three to five independent concepts simultaneously. Although the entire brain is involved in problem solving, the manner in which long-term memory is activated, as well as the type of memories "uploaded" for manipulation, can entail different processes. The recall can be deliberate, or spontaneous, or a combination of both. Some memories are emotional, some analytical, or again, a combination. An "aha! moment" of insight might be the case of a spontaneous emotional "click" in the short-term memory, which elevates a new combination of concepts to one's level of consciousness. And this is just one of many cases of how ideas are created. Therefore, to increase our chances for a creative solution, it is important in practical problem solving to engage various modes of analysis and generate different types of experiences.

Research also hints at the visual nature of insights, because our visual memory allows us to perceive more than three to five independent items. On the other hand, it can be difficult to see abstract concepts or things that are "not in the picture." For that reason, one should develop skills and train to visualize ideas. For example, multiple studies show that chess grandmasters visualize entire chessboard patterns, not individual moves or chess pieces [43].

Research also shows that IQ and general intelligence have a limited correlation with creativity, especially when the latter is defined as *divergent thinking*. Introduced by psychologist J. P. Guilford, the concept of divergent thinking describes one's ability to generate a large number of different ideas. Opposite to it is *convergent thinking*, which is the ability to focus and select a few ideas from a larger existing number of ideas. Multiple studies have demonstrated that divergent thinking is a useful skill, but by itself it is insufficient to produce highly creative outcomes. Moreover, in normal situations, one's fluency in divergent thinking only generates a *perception* of creativity among other people, not the results.

The *threshold* theory of creativity, supported by empirical data, says that at relatively high levels of IQ, with IQ above 120, the relationship between creativity and IQ is very weak [44]. Also, it seems to be domain specific, varying widely between, for example, architects, military officers, and mathematicians. In any case, relying on IQ alone will not help identify creative individuals [45].

One of the clearest examples of this notion was a massive experiment undertaken by Professor Terman in California between 1925 and 1959. During the study, researchers identified and tracked children with high IQ, hoping that these kids would become future leaders of the generation. Contrary to the expectations, and when compared to their peers, the high-IQ individuals did not distinguish themselves in their adult lives. (Moreover, the study missed two future Nobel laureates, including William Shockley. Following the outcome of the study, it was said, "No matter what measure of IQ is chosen, we would exclude 70% of our most creative children if IQ alone were used in identifying giftedness" [46]. Studies of the relationship between psychological personality traits and creativity produced somewhat predictable results as well. Among the so-called Big Five—openness to new experiences, conscientiousness, extraversion, agreeableness, and neuroticism—openness seems to be the most important factor. It is correlated to what is considered to be artistic, scientific, and everyday creativity. Conscientiousness is more relevant to creative scientists, while extraversion to "everyday creatives." For example, it can be hard to become a creative salesperson without having the ability to relate to people easily.

Since successful invention and innovation have a strong connection to creative outcomes, rather than performance in solving predefined puzzles, we believe one particular school of psychological thought deserves our close attention. It is best represented by Mihaly Csikszentmihalyi, the former head of the Department of Psychology at the University of Chicago and the author of many scientific papers and books on creativity, including *Flow, Creativity, and Optimal Experience: Psychological Studies of Flow in Consciousness.* In response to claims that creativity is a set of personal traits, he writes:

> If creativity is to retain a useful meaning, it must refer to a process that results in an idea or product that is recognized and adopted by others. Originality, freshness of perspective, divergent-thinking ability are all well and good in their own right, as desirable personal traits. But without some form of public recognition they do not constitute creativity. In fact, one might argue that such traits are not even necessary for creative accomplishment. [47, pp. 195–206]

Csikszentmihalyi and his colleagues interviewed hundreds of individuals involved in high-performance activities: scientists, inventors, artists, surgeons, rock climbers, chess players, and dancers. They found that the psychological state in which the subjects experienced a self-absorbed involvement in their work was remarkably consistent across a wide variety of activities. Csikszentmihalyi called the state *Flow*, which can be characterized as follows:

- Intense and focused concentration on what one is doing in the present moment
- Merging of action and awareness
- Loss of reflective self-consciousness (i.e., awareness of oneself as a social actor)
- A sense that one can control one's actions; that is, a sense that one can in principle deal with the situation because one knows how to respond to whatever happens next

- Distortion of temporal experience (typically, time seems to pass faster)
- Experience of the activity as intrinsically rewarding such that often the end goal is just an excuse for the process [48, p. 90]

It's important to note, "When in flow, the individual operates at full capacity" [48, p. 90]. Unlike a common trade-off situation—when the harder one works, the more tired he or she becomes—the condition of flow involves the highest level of performance combined with the desire to do as much work as possible. Figure 0.6 shows a diagram of such experience. According to the psychologists, flow occurs when a high level of challenge is met with high skills of the individual (upper right corner). When the skill level is low and the challenge is high, instead of flow, one is likely to experience anxiety or worry. On the other hand, when the skill level is high and the challenge is low, one is likely to feel bored or relaxed. As we can see, the highest level of performance cannot be achieved by skills or challenge alone. It is the right relationship between these parameters—one internal, another external—that allows us to produce the best results and leads to personal growth.

The history of inventions and major discoveries is ripe with cases when rival individuals or groups come "within inches" of each other to get breakthrough results. For example, the discovery of DNA by Watson and Crick in

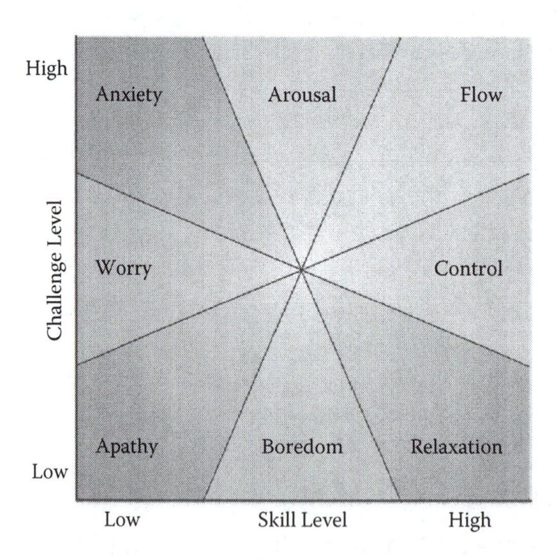

FIGURE 0.6 An illustration of the concept of *Flow* proposed by Mihaly Csikszentmihalyi. (Wikipedia. http://en.wikipedia.org/wiki/file:challenge_vs_skill.svg; Source: Mihaly Csikszentmihalyi. *Finding Flow: The Psychology of Engagement with Everyday Life*. New York: Basic Books. 1997, p. 31.)

1953 happened in close proximity to another group of researchers working on the same project. In a more recent example, the discovery of graphene that earned Geim and Novoselov the 2010 Nobel prize in physics occurred just ahead of colleagues who pursued the same line of research. Either friendly or rivalrous, the race to discovery is often a winner-takes-all competition. Small advantages, including one's conscious efforts to perform at the top of his or her abilities, can lead to a huge difference in outcome.

This threshold, nonlinear behavior of a system from slow ramp-up to wild success is even truer where inventions are concerned. For example, a one-day advantage in the priority of an invention may mean years and years of vastly different patent rights. Once we recognize that invention and innovation are not singular gestalt events, but rather a number of social processes involving diverse communities of practice, finding new potential challenges and addressing them with better skills can and should be considered a flow-like activity. We know for a fact that the best inventors, like Edison and Ford, as well as other highly productive innovation managers like Bill Gates, Steve Jobs, Larry Page, Jeff Bezos, and Mark Zuckerberg, are famous for pushing themselves and talented people around them to discover and address the toughest challenges. When we think of creativity as an outcome-oriented endeavor, rather than merely a talent that one inherits at birth, one's ability to learn new skills and apply them to important challenges becomes an important advantage in real life.

There is significant evidence, including the authors' and the readers' own experiences, that some inventions occur in a flash of inspiration. Nevertheless, the decision on whether to learn creativity as a skill does not depend on our acceptance of invention as a random, unpredictable "aha!" moment. Some ideas come in a flash, some don't. As we have discussed, the skill alone does not produce a peak performance. Moreover, high skills may lead to boredom and stagnation if the challenges we encounter do not allow us to use the skills. When popular culture and psychological studies focus on the moment of arriving at the solution, they ignore the challenge that preceded that moment, which is a key component of the creative effort. Using the analogy of Einstein's space–time, creativity happens in the challenge–skill world. In contrast, accepting the common wisdom that it is impossible to learn invention is equivalent to remaining stuck in one-dimensional thinking. Flashes of genius or any other successful efforts to solve a problem are as good as the problems they solve. As Blaise Pascal said, "Chance favors only the prepared mind," and it is our purpose to prepare our minds not for entertainment, but for highly satisfying productive work at the top of our abilities.

Summary

The common misconception about creativity as something that is innate and that cannot be learned originates from the heavy cultural emphasis on the entertainment value of an invention or discovery. Highly emotional personal events shown on the movie screen or described in a book are greatly appreciated by broad audiences. People identify with key characters and their psychological traits. As a result, common narratives about creative people emphasize the moments of discovery as unique, unpredictable events that are tied to a particular person. This approach helps storytellers effectively manipulate people's emotions, but obscures the true nature of the innovation process, which is directed at a scalable outcome. We have been taught that creativity is random, and that it belongs as a trait to a particular genius, whose existence is unique and also random.

The perceived randomness of creativity leads to ad hoc educational efforts directed largely at selection of individuals perceived as having "creativity traits." Such an education process becomes a self-fulfilling prophecy, because it begets either randomness or greater chances for success for people perceived as being creative. Compared to years of teaching traditional subjects, efforts to teach creativity are short in duration and accompanied by great expectations. This approach makes them likely to fail in the mission to increase creativity, because there is no common understanding of what creativity is and how to measure it driving the process. As a result, creativity cannot be successfully learned using the prevalent approaches today, because the practice of creativity is poorly taught, even when compared with standardized education.

Finally, scientific studies offer some interesting ideas about how various aspects of creativity can be approached to improve outcomes. For example, to raise awareness of a possible solution, we need to address the constraints of short-term memory. Long-term memories need to be expanded and augmented or even neutralized with divergent thinking to deal with psychological inertia. To leverage the strengths of different creativity styles, emotional, logical, spontaneous, and deliberate modes of thinking need to be used in creative teamwork. Successful learning of creativity as a *skill* has to be coupled with real-life challenges, rather than puzzle solving. Getting people into the habit of performing to the best of their abilities, effective and engaging challenge–discovery has to be an important part of the creative process, both in learning and in practical innovation work.

The common belief that creativity is a trait, not a skill that can be taught, is a self-fulfilling prophecy because it reduces our chances for high performance on real-life invention and innovation tasks. Among smart people with good standard education, even the smallest skill or opportunity difference can lead to a much more significant difference in outcomes. Whether one competes individually or as a member of the creative crowd, believing in popular misconceptions puts one at a disadvantage. Such beliefs are a mental barrier that can be broken by changing the way we think.

In the following chapters we apply systems thinking to discover real-life challenges, and provide tools for turning these challenges into opportunities for practical, scalable innovation.

REFERENCES

1. Levy, Steven, *In the Plex: How Google Thinks, Works, and Shapes Our Lives* (New York: Simon & Schuster, 2011).
2. Gladwell, Malcolm, *Outliers* (New York: Little, Brown, and Company, 2011).
3. Acemoglu, Daron and James A. Robinson, *Why Nations Fail* (New York: Crown Publishers, 2012).
4. Schmader, Toni, Michael Johns, and Chad Forbes, An integrated process model of stereotype threat effects on performance, *Psychological Review* 115, 2 (2008): 336–356, doi: 10.1037/0033-295X.115.2.336).
5. Brockman, John, (Ed.). *What We Believe but Cannot Prove: Today's Leading Thinkers on Science in the Age of Certainty* (New York: HarperCollins, 2006).
6. Moore, Don A. and Daylian M. Cain, Overconfidence and underconfidence: When and why people underestimate (and overestimate) the competition, *Organizational Behavior and Human Decision Processes* 103, 2 (2007): 197–213.
7. Commercial features of placebo and therapeutic efficacy, *JAMA* 299, 9 (2008): 1016–1017, doi:10.1001/jama.299.9.1016.
8. Hitting the Spot, *The Economist*, January 17, 2008, http://www.economist.com/node/10530119?story_id=10530119
9. Mankiw, N. Gregory, *Principles of Economics* (Mason, OH: South-Western Cengage Learning, 2012).
10. Congdon, William J. et al., *Policy and Choice: Public Finance through the Lens of Behavioral Economics* (Washington, DC: Brookings Institution Press, 2011).
11. Petroski, Henry, *Essential Engineer: Why Science Alone Will Not Solve Our Global Problems* (New York: Alfred A. Knopf, 2010), 155.
12. Reinschmidt, Kenneth F., Catapults of yore, *Science* 304, 5675 (2004): 1247, doi:10.1126/science.304.5675.1247a.
13. Vincenti, Walter, *What Engineers Know and How They Know It: Analytical Studies from Aeronautical History* (Baltimore, MD: Johns Hopkins University Press, 1990), 7.
14. US Bureau of Labor Statistics, *100 Years of U.S. Consumer Spending: Data for the Nation, New York City, and Boston* (Washington, DC: US Department of Labor, 2006).
15. Yahoo! selects Google as its default search engine provider, News from Google, June 26, 2000, http://googlepress.blogspot.com/2000/06/yahoo-selects-google-as-its-default.html

16. Sullivan, Danny, Yahoo renews with Google, changes results, Search Engine Watch, October 8, 2002, http://searchenginewatch.com/article/2064762/Yahoo-Renews-With-Google-Changes-Results
17. Spain's Solar Deals on Edge of Bankruptcy as Subsidies Founder, *Bloomberg News*, October 18, 2010. http://www.bloomberg.com/news/2010-10-18/spanish-solar-projects-on-brink-of-bankruptcy-as-subsidy-policies-founder.html
18. Mills, Mark P., Searching for a free lunch, finding technology, Forbes.com, September 10, 2010, http://www.forbes.com/2010/09/10/google-apple-amazon-markets-adam-smith-mokyr.html
19. Heckman, James J. and Dimitriy V. Masterov, The productivity argument for investing in young children. 2007. NBER Working Paper 13016.
20. Arthur, Brian, *Nature of Technology* (New York: Free Press, 2009).
21. Davies, Glyn, *A History of Money: From Ancient Times to the Present Day* (Cardiff: University of Wales Press, 2002).
22. Eesley, Charles. Technology and Entrepreneurship (ENGR 145), Lecture 7. Brainstorming. http://www.youtube.com/watch?V=0B69rnynnCA&list=ECF6C0319C607DEDC1
23. Larson, Kevin, The science of word recognition, Microsoft, July 2004, http://www.microsoft.com/typography/ctfonts/wordrecognition.aspx
24. Moskowitz, Howard R. and Alex Gofman, *Selling Blue Elephants: How to Make Great Products That People Want BEFORE They Even Know They Want Them* (Upper Saddle River, NJ: Prentice Hall, 2007).
25. Basulto, Dominic, Leonardo da Vinci's innovation code, Big Think, January 2, 2008, http://bigthink.com/endless-innovation/leonardo-da-vincis-innovation-code
26. McClellan, James E. III and Harold Dorn, *Science and Technology in World History*, 2nd ed. (Baltimore, MD: The Johns Hopkins University Press, 2006).
27. Hawking, Stephen and Leonard Mlodinow, *The Grand Design* (New York: Bantam, 2012).
28. Ludwig, Sean, Study: Google+ and Facebook use same ad-placement playbook, *VentureBeat*, August 16, 2011, http://venturebeat.com/2011/08/16/study-google-and-facebook-using-same-ad-placement-playbook/
29. Winograd, Terry, *Bringing Design to Software* (Addison-Wesley, 1996). http://hci.stanford.edu/publications/bds/2p-star.html
30. Brown, John S. and Paul Duguid, Mysteries of the region: Knowledge dynamics in Silicon Valley, in *The Silicon Valley Edge*, W. Miller et al., Eds. (Palo Alto, CA: Stanford University Press, 2000).
31. Brown, John S. and Paul Duguid, *The Social Life of Information* (Boston: Harvard Business School Press, 2002).
32. Markoff, John, Skilled work without the worker, *New York Times*, August 18, 2012, http://www.nytimes.com/2012/08/19/business/new-wave-of-adept-robots-is-changing-global-industry.html
33. Johnson, D. Gale, Agriculture and the wealth of nations, *American Economic Review* 87, 2 (1997): 1–12.

34. Organisation for Economic Co-operation and Development (OECD), Innovation in science, technology and industry: OECD work on patent statistics, http://www.oecd.org/innovation/innovationinsciencetechnologyandindustry/oecd-workonpatentstatistics.htm
35. Top 10 inventors of all time, Biography Online, http://www.biographyonline.net/scientists/top-10-inventors.html
36. Drucker, Peter F., *Innovation and Entrepreneurship: Practice and Principles* (New York: Harper & Row, 1985).
37. Clendening Logan, *Source Book of Medical History* (Mineola, NY: Dover Publications, 1960).
38. Ventura, Hector O., Giovanni Battista Morgagni and the foundation of modern medicine, *Clinical Cardiology* 23 (2000): 792–794.
39. Kahneman, Daniel, *Thinking Fast and Slow* (New York: Farrar, Straus, and Giroux, 2011).
40. Nisbett, Richard E. in *The Silicon Valley Edge*, C. M. Lee et al. Eds. (Palo Alto, CA: Stanford University Press, 2000).
41. Hennessey, Beth A. and Teresa M. Amabile, Creativity, *Annual Review of Psychology* 61 (2010): 569–598, doi: 10.1146/annurev.psych.093008.100416.
42. Dietrich, Arne and Riam Kanso, A review of EEG, ERP, and neuroimaging studies of creativity and insight, *Psychological Bulletin* 136, 5 (2010): 822–848, doi: 10.1037/a0019749.
43. Reingold, Eyal M., Neil Charness, Mare Pomplun, and Dave M. Stampe, Visual span in expert chess players: Evidence from eye movements, *Psychological Science* 12, 1 (2001): 48–55, http://www.ncbi.nlm.nih.gov/pubmed/11294228.
44. Sternberg, Robert J., ed., *Handbook of Creativity* (Cambridge: Cambridge University Press, 1999), 47.
45. Preckel, Franzis, Heinz Holling, and Michaela Wiese, Relationship of intelligence and creativity in gifted and non-gifted students: An investigation of threshold theory, *Personality and Individual Differences* 40, 1 (2006): 159–170, http://www.sciencedirect.com/science/article/pii/S0191886905002345.
46. Batey, Mark and Adrian Furman, Creativity, intelligence, and personality: A critical review of the scattered literature, *Genetic, Social, and General Psychology Monographs* 132, 4 (2006): 355–429.
47. Csikszentmihalyi, Mihaly, Implications of a systems perspective for the study of creativity, in *Handbook of Creativity*, Robert J. Sternberg, Ed. (Cambridge: Cambridge University Press, 1998).
48. Nakamura, Jeanne and Mihaly Csikszentmihalyi, Flow theory and research, in *Handbook of Positive Psychology*, C. R. Snyder and Shane J. Lopez, Eds. (Oxford: Oxford University Press, 2009).

Acknowledgments

We would like to thank wholeheartedly our "creative crowd"—all those who have contributed their ideas, suggestions, and support in making this book a reality. We have been learning, practicing, teaching, and writing on invention and innovation topics for many years, and we are so grateful for you helping us along each step of the way.

Thank you to (in alphabetical order by first name): Aditya Sharma, Andre Lamothe, Arjun Gupta, Doug Gilbert, Grace Hsia, Hal Louchhem, Joe Beyers, John Kelley, John Marshall, Joshua Rosenberg, Justo Hidalgo, Laurenz Laubinger, Marco dePolo, Mark Hoffberg, Michael Etelzon, Michael Irwin, Paul Henderson, Sagy Mintz, Peter Verdonk, Ron Laurie, Rosie Salcedo, Silvia Ramos, and Vladimir Petrov for the extensive feedback regarding the book's content and applications.

Our special thanks to Lev Pisarsky, who took the time to review our writing in detail and think through the entire book. Lev, you are awesome! A big appreciation for John Halloran for much the same, coupled with innumerable discussions on the topic of innovation itself and the prospect of systematizing and teaching it.

Thank you to all our students from Stanford and the University of Michigan. We thoroughly enjoyed teaching and learning alongside you throughout the years! We appreciate your enthusiastic participation and putting theory into practice, testing it out in your studies and jobs. We wish you all the success in your innovation endeavors!

Finally, we thank our families: Sofia, Emil, Vladimir, Alice, Natalia, Paul, Jonathan, Ari, Ella, Noa, and most of all Nicole and Roni. Thank you for all your suggestions and unending support, for tolerating (and engaging in) our endless stories about innovation, and the never-ending, late-night Skype sessions that it took to complete this book. We appreciate your confidence in us and encouraging us to pursue our goal in finally organizing our ideas and practices in writing, and publishing them.

Introduction to the Model

With the Prologue serving as an appetizer, we move to the book's main course—a systematic approach to creating inventions and scalable innovations. We start it with a bare-bones system model.

The economy, and indeed everything, seems to be built on a very basic premise—that goods and services will be exchanged. The exchange involves mass, energy, and/or information. These units of exchange have to originate somewhere—let's call it the *Source*—and used or transformed somewhere—let's call it the *Tool*. Mass, energy, and information have to traverse space and time—let's call the path(s) the *Distribution*. In order for the Tool to reliably receive and process mass, energy, and/or information, they have to be appropriately packaged—let's call it generically the *Packaged Payload*. Let us also introduce a *Control* element to ensure a robust, sustainable, and scalable interaction between these elements. These are the five essential, functional elements of any system. Any one of the system elements can be a system in its own right, and any system can become an element in another system. This is all.

Well, almost. Systems typically evolve along generic S curves, from inception, to explosive growth, to stagnant plateaus, until they are replaced by new systems following similar growth curves. Certain system configurations correspond to distinct places on the S curve.

Through the system lens outlined above, one can analyze and understand the function of any system, and also link the analysis to predictions about the fate of systems based on the stages of their evolution along the S curve. In this framework, typical innovation mistakes simply become system-level misfits.

This approach allows inventors, investors, and policymakers to predict and more effectively influence system development, avoiding costly mistakes, while accelerating the pace and the success rate of innovation.

In Section I we introduce the model and its core elements: Source, Tool, Distribution, Packaged Payload, and Control. We show how to make connections between the model and real-life technology solutions, including their reflection in patents. Our goal is to break through the clutter of professional jargon and buzzwords, and to solve the trade-off between the technological or business complexity and the difficulty of understanding it. The recursive nature of our model preserves the real-life complexity necessary to make rational choices but facilitates understanding by using a small number of "moving parts." In short, we show how to address the complexities of modern technologies without sacrificing the practical understanding of how they work, what they accomplish, and how they can be improved.

Section II discusses methods for navigating between system levels. We introduce tools for thinking outside the box and consider why some inventions become luckier than others. Because best innovators manage to combine great high-level vision with focused attention to detail, our goal is to offer imagination development techniques that break the trade-off between problem definitions that are too broad and too narrow.

We take into account that psychologically, we tend to focus narrowly on what is available to our immediate perception. Daniel Kahneman, a Nobel laureate who did pioneering work on the psychology of human judgment, calls this effect WYSIATI—what you see is all there is [1]. That is, "jumping to conclusions on the basis of limited evidence" is a core feature of human intuitive thinking.

To overcome the barriers presented by WYSIATI, we promote divergent thinking by systematically introducing new, multilevel perspectives. We introduce efficient tools for stretch-testing the "normal" in various dimensions—space, time, and resources—which helps to expose soft and hard barriers. The process of exploration further leads to an improvement in the quality of convergent thinking because it shows how to rationally focus on specific constraints that prevent us from making a breakthrough.

Finally, Section III deals with system dynamics: how the elements evolve along the S curve, creating space for invention and scalable innovation. We illustrate the stages and patterns of system evolution with case studies from multiple industries and technology domains. Knowing the patterns helps the reader to identify high-quality opportunities, solve problems, and fine-tune the timing of innovation.

With the understanding that efforts to produce innovations generate multiple alternative solutions, we consider two contrasting, successful strategies for introducing a novel product or service into the market. We learn that the success or failure of a particular strategy does not necessarily depend on the ingenuity of any particular invention or its immediate benefits to the consumer. Rather, the solution's *fit into the rest of the system* can determine its innovation "luck."

Using the Tech Battle technique, we explore different ways to compare innovation alternatives, separate hype from reality, and discover potential opportunities for improvements. We emphasize that the time dimension in the context of innovation is fundamentally different from the time dimension of routine processes we live through day after day, year after year. Even a simple calculation shows that our system-based approach passes the Big Bang Innovation test because it provides a 10X improvement in the number and quality of generated ideas. That is, before practicing this approach, we used to suffer from the WYSIATI bias, often focusing on only one of the elements, typically, the *Tool*. Here we provide a model for systematically exploring all five elements at two or more system levels (5 × 2 = 10). On top of that, we consider alternative implementations of the elements at different stages of system evolution, which further increases the scope of idea exploration along an S curve. And although predicting the future of innovation is hard, we believe that the concepts and tools introduced in this book can help our readers become better innovators.

REFERENCE

1. Kahneman, Daniel, *Thinking Fast and Slow* (New York: Farrar, Straus and Giroux, 2011).

About the Authors

Eugene Shteyn teaches Principles of Invention and Innovation, The Patent Paradox, The Greatest Innovations of Silicon Valley, Model-based Invention and Innovation at Stanford University Continuing Studies Program. His innovation work, first as a principal scientist at Philips Research, then as a director of IP licensing at Hewlett-Packard, is embodied in high tech products and represented in industry standards such as DVD, UPnP, and DLNA. Eugene holds 28 US patents and is a named inventor on more than fifty patents pending (in corrected media, nanotechnology, system architecture, high-performance networking, superconductivity, digital media, medical systems, Internet services, and other areas). He also founded Invention Spring, LLC, an invention development and innovation management consulting company whose clients include Fortune 500 companies (Apple, IBM, Roche) as well as startups (Ambature, Instaply, mon.ki and many others). Besides innovation, Eugene's interests include classical Japanese haiku; since 2006, he has translated into Russian over one thousand haiku by a 19th century poet Kobayashi Issa.

Max Shtein is an associate professor at the University of Michigan, Ann Arbor, teaching courses that include quantum mechanics, solid state physics, engineering design, organic electronics, innovation, and several others; he also directs a research group working on energy conversion in novel materials. Shtein earned his PhD from Princeton University in 2004 and BS from the University of California, Berkeley in 1998, both in Chemical Engineering, specializing in novel organic semiconductor physics, device design, and processing. His work and patents are used in the development of organic electronic devices (OLEDs, solar cells, transistors). Max Shtein's professional awards include the Materials Research Society Graduate Student Gold Medal Award, the Newport Award for Excellence and Leadership in Photonics and Optoelectronics, the Presidential Early Career Award for Scientists and Engineers, the Department Achievement Award, the Holt Award for Excellence in Teaching, and the College of Engineering Vulcans Prize for Excellence in Education.

Section I

Systems

> The supreme goal of all theory is to make the irreducible basic elements as simple and as few as possible without having to surrender the adequate representation of a single datum of experience.
>
> **—Albert Einstein**

> Dr. Hoenikker used to say that any scientist who couldn't explain to an eight-year-old what he was doing was a charlatan.
>
> **—Kurt Vonnegut,** ***Cat's Cradle***

In Section I we introduce a system model that explains existing inventions, technologies, and patents. We also show how to use the model for developing new ideas.

1 Invention

An Attempt to Improve the World

We begin with a simple children's poem that captures the trial-and-error nature of the invention process. In "Invention" [1], Shel Silverstein shows a boy who just made a creative breakthrough.* Excited, he shouts to the world that he invented a light bulb that plugs into the sun.

At first, the idea seems perfect: The sun has infinite energy and the new bulb converts it into an everlasting light. But, suddenly, the young inventor discovers a major flaw in his solution (Figure 1.1)—"The cord ain't long enough."

How disappointing! The breakthrough no longer looks promising and the boy seems to accept his failure to create a technological revolution.

But let's imagine for a moment that we have indeed found a way to produce the missing cord. After all, we can think of a sun ray as a 93 million-mile cord that delivers solar energy to earth. We may also imagine using a special fiber-optic cable or a satellite with mirrors, Would these new "cords" make the invention work?

Not necessarily. To make the invention useful to people, we are still missing one key element. That is, without some kind of a switch or a dimmer, the light bulb would shine incessantly, even when not needed. As a result, this invention could turn out to be annoying rather than useful. To be sure, the inventor's original intention wasn't to turn a cozy, well-lit bedroom into a 24 by 7 sleep deprivation torture chamber.

Had the inventor pursued his idea further and procured the missing cord, he would soon discover the lack of a light-control mechanism, which would further delay his innovation efforts. Even worse, his competition could invent and patent the light-control mechanism ahead of him. As a result, they would build a better lighting system and use their patents to prevent him from making and selling his solution. To summarize, the time and the resources the inventor invested into building the extra-long cord would go

* This seven-line poem can be found easily on the web by using the keywords: Shel Silverstein Invention poem (e.g., http://goo.gl/mnnfl).

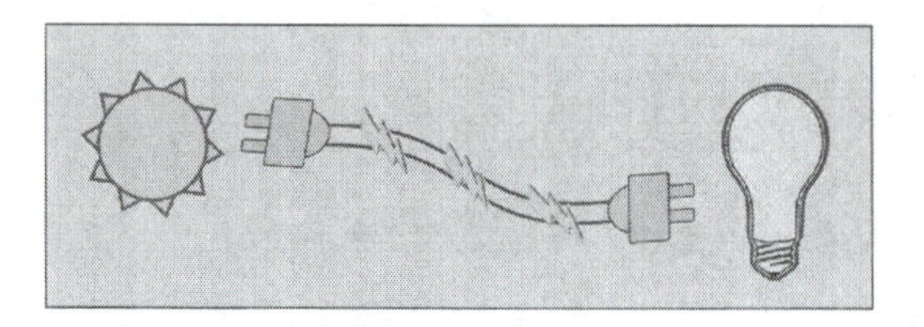

FIGURE 1.1 "The cord ain't long enough."

to waste, because he focused on inventing only certain parts of the system in isolation, instead of considering the system as a whole.

What is the whole system and what are its parts? How can we define them? Is it possible to avoid similar invention mistakes in the future?

2 Understanding Inventions
A Brief Introduction to the System Model

Let's describe the key elements of Silverstein's "invention" in more general terms, using the elements' functions rather than their implementation as a guide [2]* (Figure 2.1).

The main function of the light bulb is to turn a dark room into a comfortably lit one. As long as the bulb produces the desired outcome—the light—we can consider the purpose of the whole system fulfilled. We are going to call such an element the *Tool.*

In our model (Figure 2.2), the Tool is responsible for producing the main useful output of the system as a whole. In the poem, the light bulb is an instance of the Tool. Another possible instance of the Tool would be a special mirror that directs light into a room. On a clear night, even the Moon could be the Tool because it could reflect enough sunlight to illuminate objects in a dark room. To identify the Tool(s), we focus on entities that deliver desired system-level outcomes.

The second key object mentioned in Silverstein's invention is the sun. It produces the energy essential for our light to shine. We are going to call this element the *Source.* Its main function is to produce ingredients for one or more Tools within the system. For example, to power additional bulbs, we could have connected them to the Sun, an instance of the Source. Other potential Sources of energy would be a battery, a power station, or a solar panel. Without a Source, the Tool will not have anything to work with.

The third important object in Silverstein's poem is the cord. Its functional role is to deliver energy from the Source to the Tool. We are going to call such an element the *Distribution.* The cord is an instance of the Distribution. As we saw earlier, other instances of the Distribution could be a fiber-optic cable or a satellite with mirrors. Distribution parts, or *Routes,* connect single and multiple Sources to single or multiple Tools.

The fourth element in the invention is energy. It is the essential ingredient that flows through the cord (Distribution) from the sun (Source) toward

* The origins of the model can be found in Altshuller, page 123.

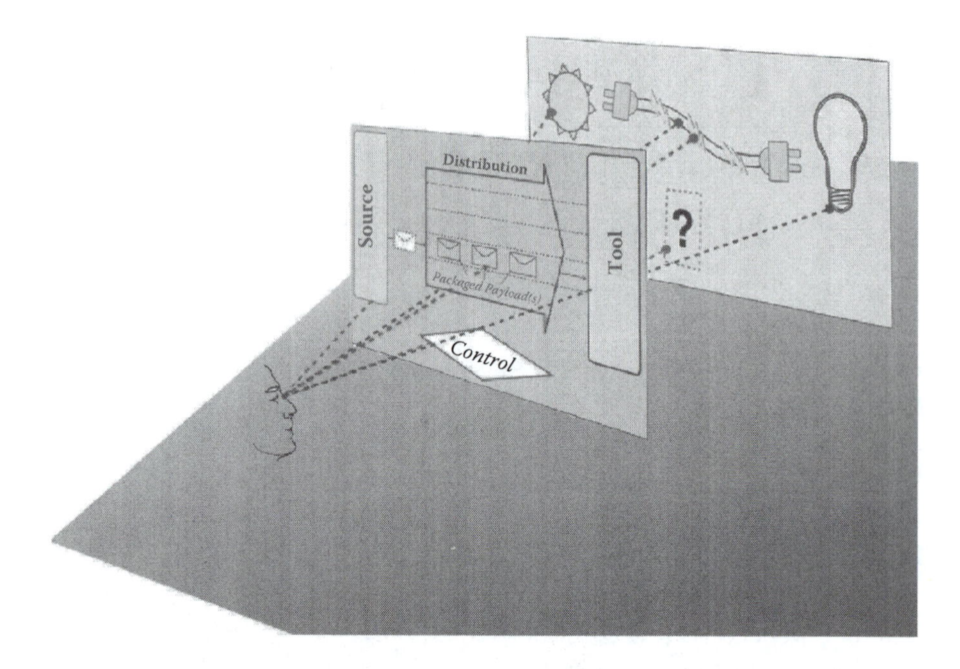

FIGURE 2.1 *Invention* as a system concept, mapped onto its physical implementation.

the light bulb (Tool). We will call it the *Packaged Payload*. We use the term *packaged* to emphasize the need for a specific form of useful energy, so that the Distribution can reliably deliver it to the Tool. Although the Packaged Payload is often hidden in the system, its form and contents determine whether the Tool(s) have something to work with.

In "Invention" the child takes it for granted that any energy from the Sun can be used to power the light bulb. But if he sent a simple ray of light to an electric bulb, the bulb would not shine. Moreover, even the wrong kind

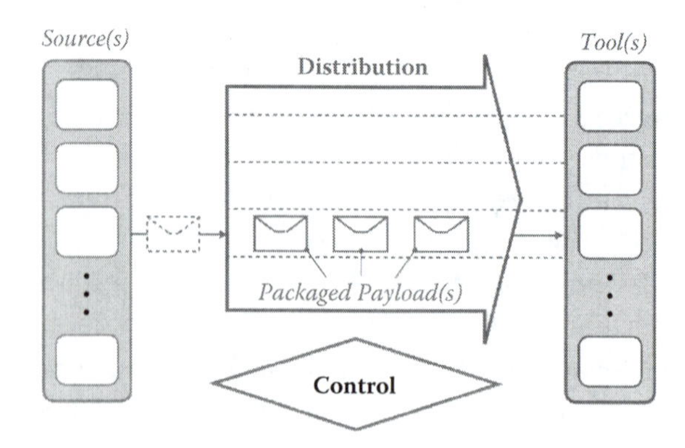

FIGURE 2.2 The system model.

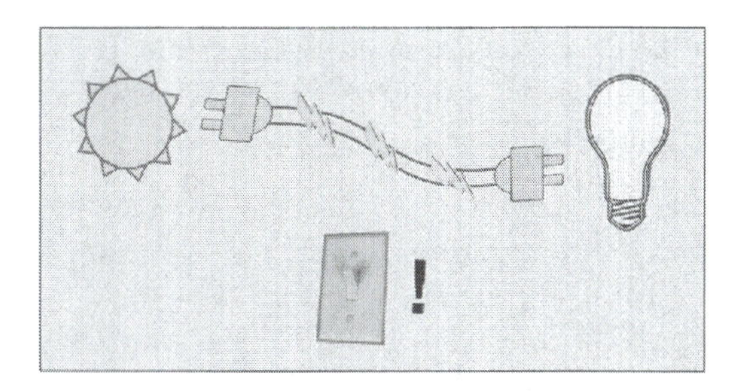

FIGURE 2.3 A working invention with all five system elements present.

of electricity, for example, with very high electric current, may damage the bulb rather than lighting it up. Just like in the Silverstein poem, the Packaged Payload is often invisible in the system. Such *invisibility* makes inventors forget about the payload and its form, and therefore they produce solutions fraught with problems. In most systems, the typical types of the Packaged Payload are energy, mass, and information.

Finally, we get to the *Control*, the functional element responsible for setting up and orchestrating interactions between various Sources, Tools, Distributions, and Packaged Payloads. In "Invention," the Control is missing, but it can be implemented as a switch that turns the light on and off (Figure 2.3). The addition of this simple element makes the lighting system as a whole adaptable to our needs. Having even a rudimentary implementation of the Control functionality dramatically changes the user's experience of the entire system.

The greater the number and variety of elements in the system, the more valuable the Control functionality becomes to the user. Let's imagine that we have not one, but a dozen light bulbs connected to the Sun. Some of the bulbs are big and take a lot of energy, while others are small and can work off a very weak current. It's the Control's job to make sure that the right amount of energy gets to the right bulb at the right time. Like a coach of a sports team, the Control makes sure that all elements coordinate their actions. Taking the sports analogy further, one could say that the Control is responsible for the team's performance as a whole.

When the elements work together as a system, the system becomes greater than the sum of its parts. That is, the Sun by itself can bring us the warmth, but it won't power the light bulb, unless we connect it to the bulb with a proper cord. Moreover, the energy intended to flow through the cord has to match the cord's physical properties and be controllable, so that we get light where and when we need it. In the simplest case, considering, selecting, and assembling known elements can result in an invention. By using the system model, we gain the ability to see beyond the immediate problem of "The cord ain't long enough." That is, we gain insights as to how to create new useful systems, find missing elements, and solve problems.

Though Silverstein's "Invention" is just an exercise in artistic imagination, the situation it describes is quite real. In fact, it is similar to Edison's early implementations of his electric system. While taking on a large-scale project to bring lights to Manhattan, the inventor built a house lighting system for J. P. Morgan, his banker and main business supporter [3, p. 8].

To generate power, Edison used a coal-fired steam engine attached to an electric dynamo. From the very beginning, the inventor knew that his energy Source would generate a lot of noise, smoke, and sparks. Not wanting to disappoint J. P. Morgan, Edison placed his power station on the edge of Morgan's property, far from the banker's home, so that the prevailing wind was blowing away from the house. Eventually, his solution drew protests from the neighbors—much like large power plants draw complaints from environmental organizations today. Just as in Silverstein's "Invention," Edison's "cord" was not long enough to get the pollution away from the residents.

If such were the problems with bringing electric lights to a single house, Edison's idea to electrify several city blocks in Manhattan seemed impossible to implement. Moreover, increasing the distance between the power station and the lights would dramatically increase wiring costs, making the entire venture unprofitable. To succeed, Edison had to think beyond the light bulb, and improve many other elements of his electric system. His inventive efforts resulted in a new parallel circuit, more efficient dynamo, an underground cable network, safety fuses, meters, sockets with switches, and many other improvements. Additionally, relying on his experience with telegraph technology, Edison brought forward the idea of using low-current DC for powering the light bulbs—an approach few electrical engineers seriously considered at the time.

In the next several chapters, we will use the system model to map out and understand various technology systems, including Edison's great inventions that brought lights to Manhattan as well as Steve Jobs' innovations, which created a revolution in mobile computing. We will show how to understand complex systems and avoid common misconceptions about inventions.

EXAMPLE 1: EDISON'S ELECTRIC GRID AS A SYSTEM

In a 2012 MIT poll surveying 1,000 young people [4], Thomas Edison was recognized as the number one innovator of all time (Figure 2.4).

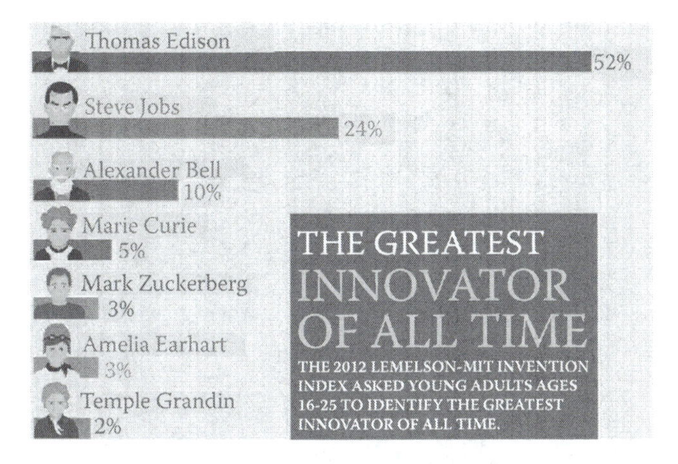

FIGURE 2.4 The diagram is courtesy the Lemelson–MIT Program. (From Lance Whitney, Edison tops Jobs as world's greatest innovator, *c|net*, January 26, 2012, http://news.cnet.com/8301-11386_3-57366904-76/edison-tops-jobs-as-worlds-greatest-innovator/.)

The article presenting the poll on the cnet.com described Edison as the creator of the light bulb. Though untrue, the idea that Edison invented the light bulb is quite popular. Even children's coloring books show Edison juggling the famous bulbs (Figure 2.5).

In reality, the light bulb itself was invented years before Edison. What he did re-invent was a high-resistance filament for the bulb. It took him almost two years and thousands of experiments to find the right material for the filament, which ended up being carbonized bamboo, no less. Because of this long quest for the magic filament material, Edison reportedly coined the now widely quoted saying, "Genius is 1 percent inspiration and 99 percent perspiration."

While it's easy to see that a long period of trial and error involves a substantial amount of perspiration, the stories about Edison's light bulb leave simple questions unanswered. What was the 1 percent inspiration? Since the light bulb was invented before him, why did Edison spend so much time looking for a particular filament, instead of using already available light bulb designs? To answer these questions and appreciate Edison's true genius, we have to look beyond the light bulb.* A system model can help us with this important task.

* In the late 1870s, English physicist Joseph Wilson Swan independently invented an incandescent light bulb with a carbon-based filament (British Patent 4933). Nevertheless, he never designed an electric system for lighting an entire section of a city.

Thomas Edison – On February 11, 1847 Thomas Alva Edison, American inventor, was born in Milan, Ohio. His inventive genius received 1,200 patents including those for incandescent electric lamp, phonograph, and key parts to the movie camera and telephone transmitter. Edison said, "Genius is 1 percent inspiration and 99 percent perspiration." His birthday is often celebrated as "Inventors Day" to encourage children to try their hands at inventing.

FIGURE 2.5 Edison showing off his light bulbs. (http://www.crayola.com/free-coloring-pages/print/thomas-edison-coloring-page/)

Imagine a construction set or a game app with five functional boxes labeled Tool, Source, Distribution, Packaged Payload, and Control. In the Tool box we find blocks that use electricity to power various devices: light bulbs, motors, refrigerators, TVs, and so on. In the Source box we find power stations, solar panels, generators, and other devices that produce electricity. To connect the Sources to multiple Tools, we pull out a set of particular wires from the Distribution box. In the Packaged Payload box we find two kinds of electricity: direct current (DC) and alternating current (AC). Finally, in the Control box we see switches, meters, relays, and other specific devices that help control the flow of electricity to the bulbs and other devices.

In Edison's case, the Source was a coal-powered steam engine attached to a dynamo machine. On September 4, 1882, the central power plant located on Pearl Street in Manhattan became the first Source in Edison's electricity distribution system. Hundreds of electric light bulbs installed in streets, offices, and businesses represented instances of the Tool. Edison's Packaged Payload, implemented as low-current DC, represented a major departure from technology solutions of the time. Most competing systems used high-current electricity, either DC or AC, to power arc lights. The higher the current, the brighter the bulbs. Why then did Edison make the switch to low-current DC electricity in the Packaged Payload?

The key to understanding Edison's decision lies in understanding his goal of creating a large-scale electric system. Instead of powering dozens of lights in a house or a single street,* Edison envisioned and implemented an electric grid that could support an area of a town, providing lighting for multiple streets, offices, department stores, and other installations. High-current solutions required thick copper wires to conduct electricity, and the greater the area one wanted to light up, the more expensive the solution became. Without making a radical change in the Packaged Payload and the Distribution, the costs of building an electric grid in Manhattan would have been prohibitive.

Unlike most electrical engineers and entrepreneurs at the time, Edison had a strong background in telegraph technology. He became a telegraph operator at the age of 15. At 27, he invented a quadruplex telegraph system and sold it to Western Union for $40,000 (approximately US$800,000 in 2012). He used the money to set up an invention lab in Menlo Park, New Jersey. Making design decisions based on his telegraph experience, Edison went against the widely accepted wisdom of the time and created a low-current DC power system. In addition, he invented an ingenious scheme for distributing electricity using parallel wires and feeder lines, rather than using existing parallel and series arrangements (Figure 2.6). This invention enabled uniform delivery of electric current to individual light bulbs and represented a major technological breakthrough for large-scale lighting installations as it eliminated high copper costs and solved many grid reliability problems. This simple, yet brilliant idea prompted Sir William Thomson, a prominent British scientist whose work in electricity and thermodynamics was instrumental in formulating core principles of modern physics, to say "No one else is Edison" [3, p. 71].

Edison directed his main creative thrust toward inventing the Distribution (parallel electric grid) and a matching Packaged Payload (low-current DC). In contrast with the (obvious) light bulb, these two critically important system elements remained invisible to the general public.

* In 1878, Pavel Yablochkov demonstrated 64 high-voltage arc lights illuminating a half-mile stretch of an avenue in Paris, France. His patents were licensed by several businessmen from different countries, but the technology did not become a commercial success due to high costs. This case highlights the difference between invention and innovation: Even a working implementation of an invention does not always translate into a scaled up (widely deployed) system.

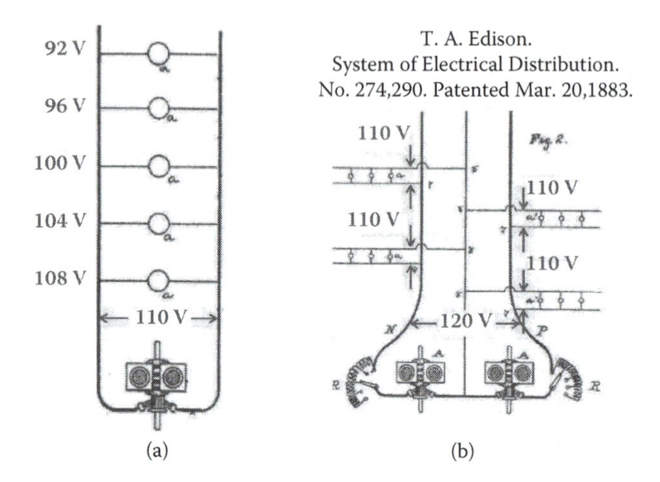

FIGURE 2.6 (a) A schematic of an ordinary parallel electricity distribution system. The voltage decreased with distance away from the electricity generator, causing the bulbs to glow less brightly, or requiring the use of thicker (and thus more expensive) wires. (b) A more sophisticated version of the parallel electricity distribution system by Edison that introduces a compensating line (ground return) that allows the use of high voltages (which reduced the amount of expensive copper wiring needed), and at the same time permits all light bulbs to continue operating, unaffected by any one burning out, for example, and also allowing for additional generators or lamp arrays to be connected more easily.

That is, insulated electric wires ran in underground street conduits, while electric current flowed through the hidden wires.

Once he solved the Distribution and Packaged Payload–related problem of high copper costs and grid reliability, Edison spent his 99 percent perspiration on discovering a suitable filament for the light bulb. Because the overall system design was based on low current, the filament had to be of high electrical resistance, so that it would burn bright enough to light a street or a large office building.

Finally, implementation of the Control elements was critical to the functionality of the grid and the success of the enterprise as a whole. Specifically, making improvements in the electric meters (protected by multiple patents) helped Edison introduce a new business model,* which allowed him to bill customers for their use of electricity. Moreover, these and other inventions enabled Control of electricity flow in an entire system. In other words, the new technology solved the problem of synchronizing power production and customer demand.

* Intel's founder, Andy Grove, is recorded as saying: "*Disruptive technology* is a misnomer. What it is, is trivial technology that screws up your business model."

> Be it known that I, Thomas A. Edison, of Menlo Park, in the county of Middlesex and State of New Jersey, have invented a new and useful Webermeter; and I do hereby declare that the following is a full and exact description of the same, reference being had to the accompanying drawings, and to the letters of reference marked thereon.
>
> In any system of furnishing to consumers electricity for light, power, or other purposes, it is desirable that means be provided which shall accurately measure the current used. It is also desirable that this measure of current should be automatically indicated and registered in a manner analogous to the registration of gas or water flow.
>
> **—US Patent 240,678. Filing date: October 7, 1880 [5]**

By examining the system as a whole and calling out its individual elements, we can recognize actual and potential innovations that remain hidden in a cursory analysis. As was the case with the light bulb, end users usually notice and invest in instances of the Tool, which are typically something they can buy to satisfy their needs. It's no wonder that manufacturers spend so much time making the Tools look attractive, especially in consumer-oriented markets! Nevertheless, all system elements (Source, Tool, Distribution, Packaged Payload, Control) should be present to deliver the desired outcome. Confusing the Tool's performance with the performance of the system as a whole is a common invention mistake. Edison was great at selling the benefits of his light bulbs to the public. At the same time, he worked hard on inventing improvements for all the other system elements that ultimately made the light bulbs so useful.

In the next section we will use system analysis to map out and understand the work of Steve Jobs, an innovator whose efforts enabled the incredible success of the iPod, iPad, and iPhone.

EXAMPLE 2: THE SYSTEM BEHIND THE IPHONE

On November 14, 2011, in Washington, DC, the United States Patent Office announced an invention exhibit dedicated to Steve Jobs (1955–2011). The exhibit features over three hundred patents granted to the inventor from 1983 to 2011 [6].

The patents are displayed on human-sized iPhone cutouts arranged in a two-sided row. Several cutouts show pictures of Steve Jobs, from the time when he was a young entrepreneur until the days when he was widely recognized as one of the greatest innovators of all time (Figure 2.7).

UPSTO's exhibit, "The Patents and Trademarks of Steve Jobs: Art and Technology That Changed the World," is now showing at the Smithsonian's Ripely Center on the National Mall in Washington, D.C.

FIGURE 2.7 USPTO exhibit of Steve Jobs' patents. (From Richard Maulsby, Start-of-summer updates for inventors, *Inventor's Eye*, June 2012, http://www.uspto.gov/inventors/independent/eye/201206/updates.jsp.)

Most people see Steve Jobs as an incredibly creative person, who spent most of his time selecting and developing great designs for Apple products—Macs, iPods, iPhones, iPads, and others. The patent exhibit seems to confirm this perception. Out of 323 patents on which he is a named inventor, most are design patents that cover product aesthetics rather than technology solutions [7]. We would not be surprised if future electronic coloring books for children show Steve Jobs juggling iPhones the same way Edison juggled light bulbs.

Vaclav Smil, an author and technology historian who in 2010 was named one of the world's top global thinkers by *Foreign Policy* magazine, while giving Steve Jobs credit for his sleek designs, noted that as an innovator he was no Thomas Edison. Smil argued that, unlike Edison's inventions spanning multiple parts of the now ubiquitous electricity distribution system, "the world without iPhone or iPad would be perfectly fine" [8].

We cannot know whether the world would prosper or suffer without Steve Jobs' inventions. But we do know for sure that psychologically people tend to focus narrowly on what is available to their immediate perception. Daniel Kahneman, a Nobel laureate who did pioneering work on the psychology of human judgment, calls this effect WYSIATI—what you see is all there is [9]. That is, jumping to conclusions on the basis of limited evidence is a core feature of our intuitive thinking.

Today, some one hundred years after Edison's seminal innovations were introduced, the majority of people still believe his greatest achievement was the invention of the light bulb. Therefore, it is no surprise that the novel, ever-present iPhone dominates discussions about Steve Jobs' achievements. As we discussed earlier, it is a common mistake to confuse the performance of the *Tool* with the workings of the system as a whole. Accordingly, to understand the potential impact of the technology revolution started by Steve Jobs, we need to escape WYSIATI and look beyond the Tool.*

Supporting the iPhone: The System Elements in Action

To simplify our initial task, let's consider the iPhone separately from other Apple devices, such as the iPad and the iPod. Once we understand the role of the iPhone within the system, it will be easier for us to add them in later.

Before we map the iPhone onto the system model, we need to appreciate the fact that there isn't just one iPhone-like device in the world, but millions, and potentially billions of them. Similar to Edison's electric grid, where the goal was to bring lots and lots of light bulbs into the city, the goal of Jobs' system was to enable communications, information access, and mobile computing for lots and lots of iPhones. Therefore, our analysis must include changes within a system that spans innovations beyond our immediate experience with a particular device.

Even a brief survey of the history of iPhone development shows that to make the device work in the real world, Jobs and his team spent enormous effort on things other than industrial design of the gadget itself. That is, Steve Jobs managed to recruit a coalition of diverse players from multiple industries: Google (cloud computing and applications), AT&T (wireless infrastructure), Hollywood, Inc. (content), Samsung (hardware and semiconductor chips), software developers (iPhone apps), Hon Hai Precision Industry Co. (electronics component manufacturer), and many others. Simply put, without this new infrastructure, the most wonderfully designed iPhone would fail in the market.

Like Edison, Jobs was great at selling the public on a novel, shiny gadget. Though its name—iPhone—sounded similar to that of a regular phone, the device became a Trojan horse that brought mobile applications and intense data exchange into everyday use. We want to emphasize that a major change in the Packaged Payload is an important indicator

* Jobs said in one of his interviews that many people mistakenly believe that "design" is how something looks. "Design is how it works." In this context, we understand that the design of the iPhone hardware cannot be divorced from the software that runs it, from the network on which it works, from the content that it presents/interacts with, i.e., we must consider the design of the system as a whole, and how this particular element (iPhone) fulfills its intended purpose.

> More than seven years before Apple Inc. rolled out the iPhone, the Nokia team showed a phone with a color touch screen set above a single button. The device was shown locating a restaurant, playing a racing game and ordering lipstick.
>
> Nokia's smartphones had hit the market too early, before consumers or wireless networks were ready to make use of them. And when the iPhone emerged, Nokia failed to recognize the threat.
>
> Nokia engineers' "tear-down" reports, according to people who saw them, emphasized that the iPhone was expensive to manufacture and only worked on second-generation networks—primitive compared with Nokia's 3G technology. One report noted that the iPhone didn't come close to passing Nokia's rigorous "drop test," in which a phone is dropped five feet onto concrete from a variety of angles.
>
> Yet consumers loved the iPhone, and by 2008 Nokia executives had realized that matching Apple's slick operating system amounted to their biggest challenge.
>
> **—*Wall Street Journal*. July 18, 2012 [10]**

that the system as a whole is undergoing a major change.* Since the Packaged Payload interacts directly with all system elements, introduction of a new type of Packaged Payload affects existing implementations of Distribution, Source, Tool, and Control (Figures 2.8 and 2.9).

In Steve Jobs' system, information is the main type of Packaged Payload. What are its new instances and how do they differ from the old ones?

The old payloads were primarily voice, short message service (SMS), and email, the latter particularly popular with users of Blackberry phones from Research in Motion (RIM). Transmission of text-based data in the old cellular phone system required low bandwidth. By contrast, the new iPhone-targeting payloads bring a heavy load of video streams, audio, web pages, map information, games, pictures, and other interactive data.

* By the term *Packaged Payload* what we mean here is arbitrary data—often, streams of data—being accessed by a wide variety of applications running on a mobile device. The type of payload that was previously reserved for separate, much larger, less portable devices (plural), was now being accessed and processed by a handheld device, continuously. For a more detailed discussion of this particular type of Packaged Payload, look ahead to Chapter 19.

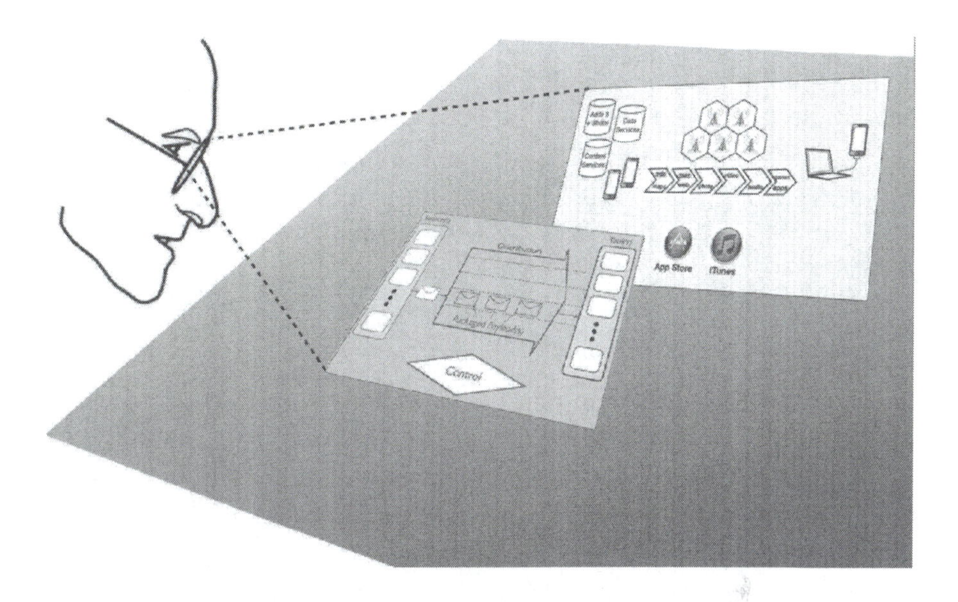

FIGURE 2.8 Mapping Steve Jobs' system.

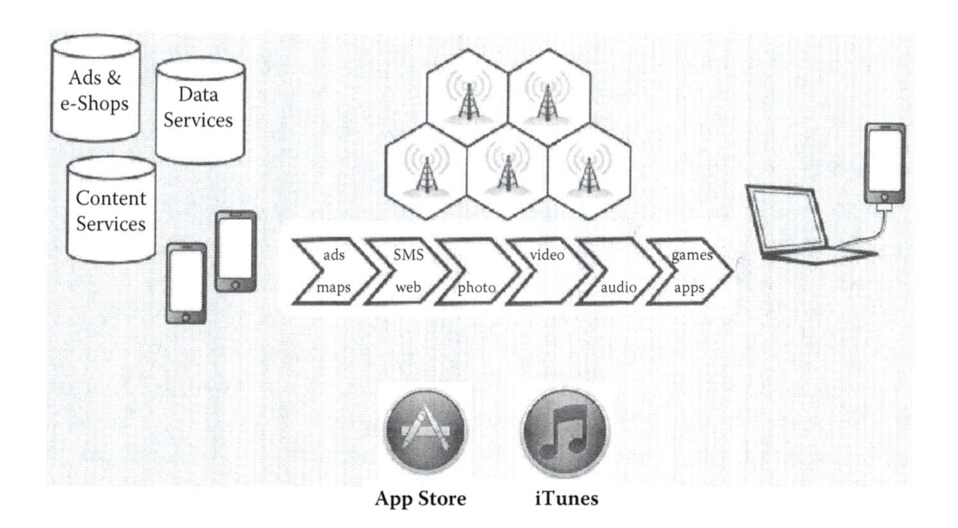

FIGURE 2.9 A diagram of the implementation layer. Element positions correspond to their system level functionality.

One particular payload, though popular on the web, is missing. Steve Jobs decided not to make the iPhone compatible with Adobe Flash. With this move, he not only avoided porting a power hog onto Apple's new device, but also sent a message to the rest of the industry that his company intended to become a de facto owner of the marketplace for mobile video.

iTunes will announce in mid-September a plan to offer movie downloads.

Lions Gate and Walt Disney will be the first companies to feature their films on the site, and the movie downloads will be priced from $9.99 to $14.99, according to the *Business Week* report.

—c|net, September 1, 2006 [11]

To comply with Apple's requirements, Google developed a special video converter, which made YouTube videos compatible with the iPhone.

Later, in Chapter 5 about system Control Points, we will consider the strategic implications of this decision. In the meantime, let's turn our attention to the Distribution.

Chief technical officer (CTO) of AT&T, John Donovan, summed up the impact of iPhone users on the existing wireless infrastructure, "There have been big changes in usage, which has forced us to throw our traditional planning models out the window" [12]. One such big change was increased demand for data bandwidth. AT&T's old edge-based network could not handle this sudden rise in the weight and intensity of the Packaged Payloads. Faced with customer complaints about network performance, the company had to accelerate deployment of 3G infrastructure.

Moreover, a new pattern of mobile communications emerged. Previously, interactions between mobile phones were largely symmetric. That is, one user with a mobile phone would place a voice call or send an SMS to another user. In either scenario, there was a mobile phone on both ends of the communication. Using system terms, we could say that instances of the Source and the Tool were largely of the same kind. Not so with the iPhone. The new device can talk not only to other phones, but also to a variety of Internet servers. For example, a lightweight request to YouTube may return a heavyweight video stream. In other words, the pattern of communications became asymmetrical, and due to a wide variety of applications and system configurations, highly unpredictable. A smarter, more robust implementation of the Distribution had to be put in place to accommodate new usage scenarios.

Although Steve Jobs did not invent high-bandwidth wireless algorithms, he forced communications companies to accelerate adoption of advanced information technologies. An increased demand for data created by millions of iPhone users and thousands of apps drove massive investment in the next generation of wireless infrastructure. An increase in mobile bandwidth set the stage for further increases in use of smartphones by consumers. Like a snowball, the increased usefulness of the iPhone led to improvements in communications, which in turn made the devices even more useful. Edison's electric grid initiated developments along the same pattern. Increases in demand for light bulbs led to large investments in the electric infrastructure, which helped introduce new electric devices and services.

On the Source side, the iPhone got out of the gate with built-in access to YouTube, maps, email, and other Internet cloud services. Since the iPhone worked over a wide-area network, content and information sources became available everywhere the network could reach. This development turned into a boon for interaction-intensive services, such as Twitter, Foursquare, Instagram, and others. Users got instant access to their social networks, making them a part of their minute-by-minute experience.

In addition to that, people started changing their shopping habits. While at a brick-and-mortar store they would use their smartphones to compare store prices with online deals [13]. Mobile ads became less annoying because they acquired local relevance. A mobile app would send GPS information to the server, which would use the location data to pick a better targeted ad. To accommodate the increase in interaction with mobile apps, the servers had to become smarter and faster. It didn't happen overnight, but the introduction of the iPhone forced content providers to be more responsive and adaptable to use patterns. Even the web, which was in existence for almost twenty years, had to serve different pages for the new type of mobile device. Remarkably, in the beginning of 2012, web access from iPhones and other Apple mobile devices exceeded that of the Mac, the more traditional form of personal computing [14]. At the end of 2012, mobile daily active Facebook users exceeded its desktop users [16].

Did Steve Jobs and his team invent the new server technologies necessary to support the iPhone? No. But they did better than that. With their technology and business strategy, they enabled other inventors and innovators to create novel Internet cloud services. Thomas Edison's introduction of an improved power station design involved the efforts of fewer than ten engineers. By contrast, building services for the iPhone required involvement of thousands of engineers, software developers, and college and high school students. To a large extent, the work of Steve Jobs and his team triggered exponential growth in mobile services, which resulted in rapid development of cloud computing.

From a system perspective, the iPhone plays the role of the Tool most of the time. The device delivers end-user experience for the communications system as a whole. Its physical design and ease of use serve this purpose well. But, unlike Edison's light bulb, it is not a single-purpose device. The iPhone provides a software and a hardware platform for a wide range of applications. Each of the applications turns the iPhone into a slightly different Tool. For example, it can work as a communicator, a GPS navigator, a gaming console, a video console, a banking terminal, a digital wallet, e-book reader, and much more. At the time of writing this book, there are more than 500,000 apps available for the device. Within any given slice of time and space, the app plays the role of a virtual Tool that delivers the system's functionality.

Later in the book we will consider implications of the iPhone's multitouch interface for the development of modern information technologies. For now, we will only mention its ability to zoom in and out of a stream of information. Similar to the personal computing revolution, which brought about the computer mouse and graphical user interface, the iPhone is a radical departure from the old ways of working with information. The hundreds of thousands of mobile apps represent just the tip of the iceberg within a system of data services they can and will provide.

For example, the new device could create a revolution in healthcare [16]. A mobile connected device like the iPhone can be easily connected to sensors, X-ray machines, blood test kits, exercise equipment, and much more. In another example, the military is increasingly turning to smartphones and tablets to enable computing and communications on the battlefield. All these applications go beyond exploiting the beautiful industrial design envisioned and implemented by Steve Jobs and his team. These developments show that the iPhone represents much more than just a nice consumer toy. Edison's low-current light bulb was the first important instance of the Tool in his town-scale electricity distribution system. Similarly, the iPhone is the first important instance of the Tool in the new large-scale information system.

The iPhone's flexibility allows it to play a role not only of the Tool, but also of the Source. That is, it can originate information, such as voice, photographs, or videos, and send it to other instances of the Tool for interaction and processing. Moreover, due to its interface and computing capabilities, the iPhone can play the role of the Control within smaller systems. For example, with a remote control app it can coordinate home audio, video, and heating, ventilation, and air conditioning (HVAC) equipment. Some inventors use it for controlling robots, from toys, to house cleaners, to drones. All these applications show the ease with which an iPhone-like device helps create new systems out of available technology building blocks. It is a strong indicator of a computing revolution pervading every aspect of modern life.

iTunes and the AppStore function as the Control in Jobs' mobile communication system. They are essential for the success of the system as a whole because they help consumers set up and manage their devices, applications,

> To manage a system effectively, you might focus on the interactions of the parts rather than their behavior taken separately.
>
> **—Russell L. Ackoff [17]**

music, video, books, contacts, software upgrades, subscriptions, backups, and more. Because Apple provides iTunes and the AppStore at no (obvious) charge to the consumer, a cursory analysis often underestimates their impact on the system as a whole.

To appreciate their true value to the company, let's conduct a thought experiment with Bob, an iPhone user, who lost his device. Although this loss would be a disappointment to Bob, he can easily replace it with a new iPhone. Since the backup of all digital contents of the device can be set up to occur automatically, the new iPhone will have all the content and applications back as soon as Bob synchronizes it with iTunes. Replacing or upgrading the hardware is a relatively painless experience.

Similarly, losing access to a Source-based service (e.g., e-books from Amazon or iBooks) is not a big problem for Bob. He can easily find another one that provides similar content or information. By continuing our experiment, we can see that when Bob loses his AT&T 3G data connection, he can quickly replace it with Wi-Fi, at least temporarily. Also, if he wishes, Bob can switch his contract to another wireless carrier and keep all his content, apps, and contacts. If Bob loses the ability to process a certain video format, he can go to the App Store and find one of a half-million applications, which for a modest fee would restore that capability.

To summarize, when Bob loses specific instances of the Source, Tool, and even Distribution, he can recover them quite easily. Not so with the Control. If Bob can't access iTunes and the App Store, he loses access to most of his content, backups, the content itself, software upgrades, and so on. Some users attempt to circumvent the iTunes ecosystem, for example, through the practice of jailbreaking, and shoulder (or enjoy) the burden of system management. Without the Control, the services responsible for organizing information flows and security of the system, Bob's iPhone becomes rather limited in its functionality. For example, when his friends download a cool new social game, he is left behind with an orphaned device. Losing iTunes or access to the App Store requires Bob to perform all operations manually. The more content and apps he has on his device, the more time he will have to spend working as a human version of Apple's Control software.

Is this situation an accident of system design? Hardly. Apple's free Control component simplifies replacement of any system element, except itself. By seamlessly coordinating Bob's care of his iPhone and its contents, iTunes and the App Store help him choose a new version of the iPhone when the time comes to replace the hardware. Furthermore, if Bob wants to buy a new tablet, he can add it to the system with minimal effort, provided the tablet is an iPad. Out of all the system elements, the Control is the stickiest.*

As we apply our model to Steve Jobs' system, we discover its key features hidden behind the ever-present iPhone. Because of this new perspective, we begin to appreciate the effort required for a successful innovation. Furthermore, instead of superficially comparing creative individuals (e.g., Steve Jobs vs. Thomas Edison) we learn how to compare the systems they created.

For example, by a simple patent count—1093 versus 323—we know that Edison made more inventions than Jobs. But through system analysis, we learn that Edison's inventions went well beyond the light bulb and spanned multiple system elements: Tool, Source, Distribution, Packaged Payload, and Control. In contrast, Steve Jobs focused on inventing one or two critical elements and developing extensive partnerships with industry players to make the system successful as a whole. Both of them succeeded in laying the foundation of a major technology and business revolution of their time.

We also know that inventions created by Edison or Jobs are not the last word in the process of innovation. In Sections II and III, we will learn in greater detail how the system model helps discover opportunities for innovation. In the next chapter, we will show how to use the system model to understand patents.

REVIEW QUESTIONS

1. In November, 2012, AT&T announced that eventually it intends to decommission its copper-line phone network and spend $14B on replacing it with broadband and high-speed wireless services.

   Name a system element affected by the announcement and outline the rest of the system [18]. Draw the system diagram and specify

---

* Later in the book we will consider how the Control works in greater detail. For now, we stress that Apple is not alone at leveraging the power of "free" Control functionality. For example, Facebook and LinkedIn provide services for managing your social connections, which you cannot transfer in any practical way to a different platform. Google, through a single account sign-in, wants you to organize your world's information through its search, browser, mapping software, email, documents, and a myriad other apps and services. The more elements that comprise the system, the more value the Control part holds within it. Even when it comes for "free."

instances of the system components. What type of Packaged Payload will the System handle: mass, energy, or information? What other companies are going to benefit from the system's growth?

2. According to MIT Technology Review, "an energy startup called Stem has developed a battery for commercial buildings that's clever enough to predict—based on the price of electricity—when to store power and when to release it. The batteries aren't just intended as backup power. Instead, in concert with software, they're part of a system designed to allow a building to use the cheapest form of power available at any given moment, whether that power comes from Stem's batteries or from the grid" [19].

   Name the system element(s) implemented by the startup and outline the rest of the system. Imagine you are the chief executive officer (CEO) of the company. Which system elements would you ask your engineers to invent or improve for the long-term benefit of the company (and its customers)? Which system elements would you ask your sales people to sell in the short term? Which system element(s) do you need to control to become a dominant force in the industry?

3. On August 13, 2012, the *New York Times* noted, "The Bill and Melinda Gates Foundation awarded $3 million to researchers at eight universities, challenging them to use recent technology to create [toilets] that needn't be connected to sewers, or to water and electricity lines, and that cost less than pennies per person a day to use" [20].

   Outline the new system in which such toilets would participate. Within the system, is the toilet an instance of the Tool, Source, Distribution, Packaged Payload, or Control? Outline the old system. What are the most important differences between the old and the new systems? Discuss with friends.

3 Understanding Patents
An Application of the System Model

Understanding patents is often difficult for nonspecialists. From the authors' personal experience, even when you are the inventor of the solution, explaining it to your patent attorney and figuring out how patent claims relate to your invention can be a challenge. As a result, years after spending their precious time and money, many inventors and entrepreneurs discover that their patents are worthless.

Furthermore, comparing patented inventions from two completely different technology areas is often considered outright impossible. As the common wisdom goes, you can't compare apples to oranges. Nevertheless, if we want to learn how to invent and innovate effectively, we need to know how to recognize, compare, and build creative solutions across multiple technology fields.

EXAMPLE 1: PLASMON WAVEGUIDE

Let us consider Claim 1 from US Patent 7,542,633 [21]:

> 1. A method comprising: guiding a plasmon signal on a first plasmon guide; selectively controlling the guided plasmon signal with a plurality of control signals, wherein at least one of the plurality of control signals includes optical electromagnetic energy guided through a waveguide that is directly coupled to the first plasmon guide.

Many people, including scientists and engineers, find this claim difficult to understand. It contains a mix of technical and legal terms that have little in common with our everyday experiences. Even when Eugene, one of the coauthors of the book, reads the claim for the first time to his Stanford University Continuous Education students—many of them do have physics, engineering, or legal backgrounds—their eyes glaze over.

Nevertheless, a small transformation can make understanding this invention a lot easier. Let's conduct a thought experiment and substitute in our mind's eye *plasmon signal* with *train*. Now, if the plasmon signal is a train, then the plasmon guide should play the role of train tracks.

So far, so good: Instead of the difficult-to-understand plasmon, we've got a much more familiar train moving along the train tracks. Working further, let's see how our train is controlled. Well, even an eight-year-old can tell us that he would need a semaphore (a device with different colored lights, e.g., green and red) to signal the train's engineer whether he should drive or stop. With the plasmon, instead of the semaphore we have *a plurality of control signals*. The implementation of the semaphore is quite different, but the functionality is essentially the same: provide control signals for the movement along the path.

Finally, how does our semaphore get the energy to power the lights? A simple investigation into the railroad technology reveals an electric wire, running along the tracks, either buried in the ground or strung on poles. Again, the electromagnetic waveguide in the invention is very different from the wire, but conceptually, it is just a separate physical route to deliver energy for the control signals. Amazingly, a rather simple substitution—we could have used a car, a road, and traffic lights—gave us the ability to understand what looked like a very complex invention.

Spotting System Elements in a Patent Claim

By now, you have probably recognized some of the system elements we discussed earlier (see Figure 3.1). The plasmon and the train play the role of the Packaged Payload. They are being transferred from the Source to the Tool, which are not mentioned in the claim. The plasmon delivers either energy or information. If necessary, we could figure out its precise role within the system by reading the rest of the patent.

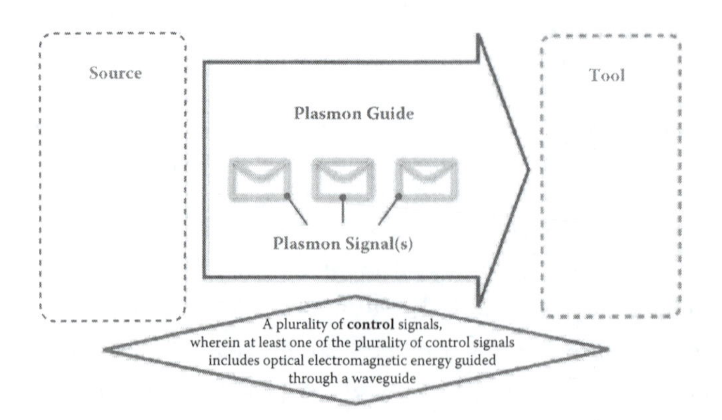

FIGURE 3.1 Guiding Plasmon Signal, US Patent 7,542,633.

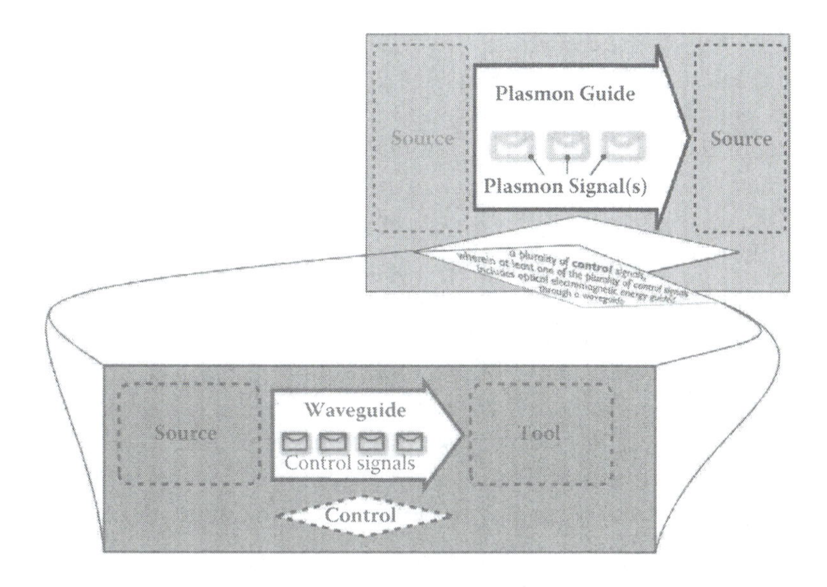

FIGURE 3.2 *Control* subsystem within a system.

The plasmon guide and the rails are both Distribution elements. Their function is to facilitate the transfer of their respective payloads from the Source to the Tool.

The Control functionality hinted at by the plurality of control signals deserves some additional attention because it involves several elements: the signals, optical electromagnetic energy, and waveguide.

The key to understanding the elements' role in the invention is seeing the Control as a subsystem from the vantage point of a higher-level system. That is, we are adopting a nested, hierarchical view of systems and elements, wherein elements themselves can become systems when inspected more closely, while systems can become elements, when viewed from a sufficient "height." In other words, the Control element now has its own Tool, Source, Distribution, Packaged Payload, and Control subelements (Figure 3.2).

The optical electromagnetic energy represents the Packaged Payload. The Tool is omitted in the claim, but it is ultimately responsible for translating the energy into physical control of the plasmon.

The waveguide acts as the Distribution because it facilitates transfer of signals from the Source omitted in the claim.

We don't see in the claim how the Control is implemented, but it's easy to figure out that something in the system has to make a decision when and how to affect the transfer of the plasmon.

Clearly, one of the difficulties in understanding patent claims lies not only in cracking the technical and legal jargon, but also in recognizing the strategic omissions. Patent lawyers and agents are trained to claim essential elements of the invention only. Often, the fewer terms that are used in the claim, the broader its potential scope of interpretation.

By applying the system model, we kill many birds with one stone. First, we eliminate the jargon and identify parts omitted in the claims. Then, we check whether the omissions were created by mistake rather than strategic intent. Finally, we systematically explore possible implementations for both present and missing elements. We can use the same technique to check patent claims for completeness and consistency, and discover workarounds when it is necessary to overcome somebody else's patent. As a result, we find opportunities for inventions that were overlooked previously. If needed, we design claim sets and patent portfolios to cover different elements and system levels.

In our next example we will consider US Patent 7,092,861, Visual Anti-Virus in a Network Control Environment, awarded to one of the coauthors (Eugene Shteyn). He made the invention in early 2000; the US Patents and Trademarks Office (USPTO) issued a patent in 2006; Facebook acquired the patent in 2011.

EXAMPLE 2: BUILDING AN INTELLIGENT REMOTE CONTROL

In February 1985, Steve Wozniak—cofounder of Apple Computers—left the company to start a new venture. He wanted to build a universal remote control device to replace all other remote controls in the house. At the time, a standard remote could control one type of device only, such as a TV or VCR. To turn the device on, off, or change channel, the user would have to press a designated button on a device-specific remote. When pressed, the button would trigger the remote control's circuitry, which in turn would send an infrared control signal to the TV or VCR. If the user needed to send a sequence of commands, he would have to push many buttons. The more devices, the more complex the sequence, the greater were the chances that the user would make a mistake by pressing a wrong button on a wrong remote. As a result, he would have to start over with no guarantee of success.

Steve Wozniak wanted to solve this problem by producing a programmable, learning remote control. The remote would learn a sequence of commands, store, and send it (with the press of a single button) to the right devices at the right time. He even invented his own programming language and built it into the remote. On April 1, 1990, Wozniak and his colleague

So typically I would turn on the TV with one remote control and maybe I'd turn on the hi-fi with another (because I had the speakers routed to the TV), and then I would turn on the satellite and then I'd press a few buttons for the channel I wanted on the satellite and I think I had to turn on my VCR to pass a signal through it—all the signals passed through it to get to the TV the way I had it hooked up. I'm pushing all these buttons on different remote controls and it was just obvious to me.

Here I am, sitting in bed, operating all of this equipment with all of these different remote controls.

It was crazy. I wanted one remote control with one button that was programmable to deal with all of the devices.

—Steve Wozniak, *iWoz* [22, p. 261]

Charles H. Van Dusen received US Patent 4,918,439, Remote Control Device. The patent claimed a remote control with an internal memory unit for storing control commands and a connection to another device, for example, a computer, for loading the commands into the remote's memory.

Although it was a technological marvel, the remote control did not become a commercial success. It was too expensive for replacing ordinary remote controls. Besides, few consumers had multiple pieces of audio-video equipment to justify the purchase of a universal controller. Additionally, consumer electronics devices at that time were too "dumb" to work well with Wozniak's programmable remote. That is, while they could receive and execute commands, they couldn't communicate back to the remote. In a simple example, if someone turned on the VCR by using the power button on the device itself or on a different remote, the new programmable remote wouldn't know that the VCR was on. Therefore, when executing a sequence of commands, for example, for scheduled recording, it would send a command to turn the VCR on. Because on most VCRs the on and off commands were the same, the VCR would turn off instead of on, and the recording would fail. What's worse, the hapless programmable remote would not learn about the failure and continue issuing preprogrammed commands to the VCR and other devices, getting the entire audio-video system into an unknown state. Even if the programmable remote was cheap enough to appeal to the average consumer, it would have trouble working with ordinary consumer electronics devices. (Later in the book, especially in Section III, we will discuss how to use the system model to anticipate and solve such system-level mismatches.) Eventually, Steve Wozniak decided to sell his remote control business and spend more time with his kids. The problem of the intelligent remote control for the home remained unsolved.*

DISCOVERING NEW OPPORTUNITIES AND ADDRESSING NEW THREATS

In the late 1990s, Eugene was working as a principal scientist at Philips Research Silicon Valley. One of his areas of interest was building networks of devices. At the time, a great number of homes and businesses were connecting to the Internet; prices for digital processors, computer memory, and software were also falling. As a consequence, and in contrast with the situation in the late 1980s, "smart" hardware was becoming much cheaper, enabling consumers and businesses to purchase more devices. Moreover, these new devices could form a network and communicate with each other easily. (This is becoming more widely known now as "the Internet of Things.") Thus, the issues that plagued Wozniak's programmable remote control just several years earlier began to disappear. In short, an opportunity emerged to implement Steve Wozniak's idea and control a group of devices using a software program.

The good news was that the new system had enough computing power and network capacity to run sophisticated programs, coordinating actions of multiple devices. The user no longer needed to push many buttons to get her audio-video equipment or heat, ventilation, and air conditioning (HVAC) system to perform coordinated actions. Similarly, a business could run new applications to coordinate office or factory equipment.

The bad news was that a software virus, or any other malicious program designed to harm the system, could run with the same ease and impact. Because the user no longer controlled every step of the command sequence, the virus would be able to exploit this attention gap. Unlike traditional computer viruses, this other type of virus could cause serious physical damage to people, equipment, buildings, and the environment. By the time the user could see that something went wrong, the damage would already be done.

These concerns materialized ten years later, when the Stuxnet virus was discovered in Iran. The virus, reportedly developed by the United States and Israel, targeted a secure industrial facility located at Natanz, allegedly used

* We give this example as an illustration that even a good invention (a simplified remote control) by a brilliant inventor can fail for reasons dictated by the higher level system, not the solution itself. An an exercise, it may be worthwhile to identify the system elements surrounding this invention, and specify which of the system elements and interactions were responsible for the failure.

for refining radioactive material into a weapons-grade form. According to the *New York Times*, "From his first months in the office, President Obama secretly ordered increasingly sophisticated attacks on the computer systems that run main nuclear enrichment facilities, significantly expanding America's first sustained use of cyberweapons" [23]. Since then, suspected government-sponsored viruses, named Flame, Duqu, and Gauss, were found on computers in the Middle East and elsewhere around the world. We can reasonably expect that the proliferation of networks of devices (i.e., the growth of the Internet of Things) will provide a fertile breeding ground for such viruses.

> The Natanz plant was hit by a newer version of the computer worm, and then another after that. The last of that series of attacks, a few weeks after Stuxnet was detected around the world, temporarily took out nearly 1,000 of the 5,000 centrifuges Iran had spinning at the time to purify uranium.
>
> Internal Obama administration estimates say Iran's nuclear program was set back by 18 months to two years, but some experts inside and outside the government are more skeptical, noting that Iran's enrichment levels have steadily recovered, giving the country enough fuel today for five or more weapons, with additional enrichment.
>
> **—*New York Times*, August 14, 2012 [23]**

USING THE SYSTEM MODEL TO SOLVE THE PROBLEM AND PATENT THE SOLUTION

Let us free up some of our mental space by forgetting for a minute how the connected devices, network, and viruses are implemented, and consider the situation from a high-level system perspective.* Here we can see that an entire programmable network of devices in one home or office would represent an instance of the Tool. Why? Because it is the function of the Tool to receive and execute a program, making sure the system performs a desired sequence of actions. That is, with the right program, the entire network of home devices can execute watching-favorite-movies, doing-dishes-and-washing-clothes-while-saving-power, or some other user scenario. Importantly, in this perspective we consider the situation from outside the house or the office.

* See Chapter 9, "The Three Magicians," in Section II for more detail on multilevel system analysis.

Therefore, we don't worry yet about how the software and hardware work together inside the house. Just like Apple's motivation with the App Store, our goal is to enable rapid scale-up, by providing programs to hundreds, thousands, and millions of individual houses and offices electronically.

How would we know what kind of programs (or apps) the Tool is capable of executing? How smart should the Tool be to run the apps? Claim 22 of the patent provides an answer: "A method of creating a customer base, the method comprising: specifying a user's inventory of equipment to a server on a data network; modeling the control of the equipment, based on one or more scripts; and storing information in the customer base, based on the user's inventory of equipment" [24].

As you can see from the claim, the server has a customer database where it keeps an inventory of user equipment. Because it would be too cumbersome, expensive, or even dangerous to run a program (the patent calls the program a *script*) on the user's actual physical equipment in the home or office, the server models the way the program controls the equipment. To discover potential conflicts, the server can model how multiple programs interact within the home. That is, even if each program runs safely on the Tool, their combination may clash for device and network resources. Because breaking the model, instead of the house, is harmless, we can run the programs without damaging user equipment. Therefore, our tests can be as rigorous and as lengthy as necessary to ensure the Tool is free from undesired side effects. Once modeling is finished, the server becomes an instance of the Source in the program distribution system. Logically, programs in such a system are instances of the Packaged Payload.

How do we identify the other elements of the system? In Claim 20 of the patent we read: "communicating with the user to facilitate the purchase of said items and/or services based on the user's equipment profile" [24].

The phrase "communicating with the user to facilitate the purchase" describes an instance of the Control. That is, by purchasing, the user controls the transfer of the Packaged Payload (items and/or services) from the Source (server) to the Tool (home/office equipment). Note that in addition to providing the programs, the system can sell other items or services related to the user's equipment. The claims omit the Distribution component because the invention does not focus on this particular aspect of the system. We can reasonably assume that the Tool has access to the Internet or some other network. As we mentioned earlier, the system, for example, at the Source, is set up to model the control of user equipment and weed out programs that cause potential breakdowns. If a program is infected with a virus, its abnormal behavior will show up when we test it on the model. (Table 3.1 summarizes the high-level system configuration.)

TABLE 3.1
A High-Level Description of System Configuration According to the US Patent 7,092,861

Packaged Payload(s)	Script or App, Description of user equipment
Tool(s)	Networked user equipment, e.g. in the home, office, factory
Source(s)	Server with a customer base of user equipment configurations and scripts
Distribution(s)	Network, e.g., the Internet
Control(s)	A service that facilitates checking, purchasing, or tranfer of scripts, equipment, etc.

To summarize, with a system-based approach we (a) identified potential problems and opportunities created by new technology developments, (b) took advantage of the opportunities and solved relevant problems, and (c) implemented a solution and developed a patent, which was likely to maintain value over its twenty-year lifespan.

Industry estimates show that only 1 percent of all patents become highly valuable [25]. While developing a patent portfolio, we can monitor how well each patent covers specific system elements. Moreover, we can find potentially valuable problems at different system levels. For example, Eugene's initial investigation into network security problems lead to problem-hunting projects joined by other researchers. They developed a number of solutions and embedded them into specialized computer chips. (US patents 6,985,845, 7,110,858, 7,257,839, 7,694,190 were later transferred to a semiconductor manufacturing company. They couldn't patent a new system-level virus itself because US patent law allows patents for useful inventions only.) Although detailed patent strategy falls outside the scope of this book, the reader should appreciate the ways in which the system model helps to cut the possible 99 percent waste in the portfolio.

In the following section, we analyze the patented antivirus solution in even greater detail. Readers not interested in software and device security topics can skip the remainder of this subsection and proceed to the next chapter.

With the basic structure of the high-level system laid out in the open, we can examine its specific security problems. Starting with the Packaged Payload, we check whether conventional antivirus methods can succeed in disinfecting software destined for a network of devices. Such methods search programs for known patterns of malicious code. For example, when we download a PC app from the Internet, our antivirus program checks the contents of the app against a database of known viruses. If it finds a match, the antivirus logic deletes or quarantines the app, preventing it from running on our computer.

Obviously, as the Stuxnet case has shown, the approach fails to detect new or unique viruses. Relying upon it would create a critical security hole in our system because it is designed to be flexible and customizable with unique apps. That is, we want users and third parties to easily add devices to the network, modify network configurations, and create programmable user scenarios. As a result, the system will have a large number of new or unique instances of the Tool. Testing them for known blocks of malicious code will not work.

Another traditional approach—direct emulation of the controller that runs the program—will not work either because the virus may target not the controller itself, but rather a device that executes commands delivered to it by the controller. For example, in the Stuxnet case, the virus spread via Windows PCs, leaving them unharmed. Instead, it targeted microprocessors embedded in uranium enrichment centrifuges. The malicious code caused them to spin erratically and break down mechanically. Similarly, in a home or office environment, a virus may cause HVAC devices to overheat, or it may unlock doors or windows while the owners are away. During all this mischief, the controller, for example a Windows PC, would suffer no harm. Emulating its performance would do little to detect the virus.

Because the old methods don't work in the new environment, we have to model the entire Tool, including all its devices, programs, and connections. Furthermore, we want to show to the user the results of a program's work before the program has a chance to affect the system. In a simple case, we provide two modules. One of them models the work of the controller, controlled devices, and their interconnections; another helps visualize the state of the model. Here's how it is reflected in Claim 1: "A system for detecting a potential virus in a control script, comprising: a modeling system that is configured to create a model of a control system, based on a network description corresponding to a control structure described by said control script, said network description comprising a combination of control and controlled devices and their interconnections, and a rendering system that is configured to provide a visual representation of the model of the control system, wherein said visual representation facilitates the detection of said potential virus to a user" [24].

The *modeling system* above represents the Source, while the *rendering system* represents the Tool. Since the Source and the Tool are functional elements, they can be implemented either on the same physical device, or on different ones. For instance, the modeling system (the Source) may run on an Internet server, while the rendering system (the Tool) may show simulation results on a personal device, such as a mobile phone or tablet. The claim omits how the Packaged Payload and Distribution work, but we can easily send an encrypted web page from the Source to the Tool over a secure Internet channel.

Additional claims and the patent specification describe various other implementation examples. As we explore the system element by element, we ensure that no security or implementation hole escapes our attention. The system model helps us develop robust solutions and protect them with strong patent coverage.

4 System Interfaces

How the Elements Work Together

In previous chapters we focused on identifying specific technical elements and mapping them onto the system model. Now, we are going to examine how elements interface with each other.

An interface between elements is a high-value innovation target because it enables the system to grow and evolve—in Lego-like fashion—through addition, removal, and replacement of its elements. For instance, power outlets in our houses make it easy for us to add electric devices, move them from room to room, replace light bulbs, and so on.

System interfaces let the ever-changing elements interact with each other, so that all of them, while working together, can reliably deliver the functionality of the system as a whole. In the house lighting example, the humble electric socket ensures that once you screw in a new light bulb and flip the power switch—let there be light—the electricity is guaranteed to flow into the bulb.

The online version of the *Merriam–Webster Dictionary* defines the word interface as the place at which independent and often unrelated systems meet and act on or communicate with each other.* Upon close consideration, the definition appears contradictory, because systems capable of interacting with each other cannot be considered independent, at least for the duration of their interaction. In other words, we can formulate this interface paradox as follows:

- on the one hand, the interface should ensure independence of parts, so that they can be easily changed, added, or removed;
- on the other hand, the interface should ensure that the parts can depend on each other, so that they work reliably together as elements of a larger system.

* Here we must once again underscore the possibility that an element of a given system is a subsystem itself. Thus, we can refer to the "interface between system elements" and the "interface between systems" interchangeably. Also important is the fact that the interface can become a major Control Point, thus becoming an element or a subsystem. That is, our analysis takes a heavily nested, recursive view of complex systems and processes.

It is a challenge for the inventor to resolve the paradox and come up with a best-of-both-worlds solution: allow the interacting elements to evolve independently *and* enable them to interact with each other seamlessly and efficiently. In this chapter we will consider examples of system interfaces and discover how they impact the rest of the system.

INTERFACE EXAMPLE 1: EDISON'S LIGHT BULB SOCKET

Over the last hundred years, the media paid an inordinate amount of attention to the invention of the light bulb. The light bulb itself has become a symbol of a creative idea. When you search the Internet for images corresponding to the term *idea*, the very first page will primarily be pictures of light bulbs. When someone describes a process of coming up with a bright idea, you can often hear, "A light bulb went on in my head." Like evening butterflies, we seem to be keenly attracted to the light bulb, both to its symbols and implements. One possible reason for the light bulb's incredible prominence in our collective mind is a thirty-five-year advertisement campaign General Electric (GE) began in the early twentieth century. After a long series of mergers and acquisitions, including the merger between Edison Electric and Thomson-Houston Company in 1892, GE ended up controlling 80 percent of the lamp market in the United States.

In 1909, rather than trying to explain to the public the technological advantages of its newest inventions, GE described the light bulb as the Sun's only rival (Figure 4.1).

Moreover, GE's light bulb ads prominently featured Edison (the greatest inventor of all time) and GE's own engineering marvel—a long-lasting bulb with a tungsten filament, which successfully replaced Edison's original invention. Mesmerized by clever pictures and bright lights, the public happily overlooked Edison's longest-lasting invention: the screw-in light bulb socket (Figure 4.2). (That is, while many people still know and use the term *Edison socket*, few think of this invention as the greatest invention by Edison.)

The history of lighting technologies shows a steady pace of innovation, going from Edison's original carbonized bamboo, to tungsten-based filaments, to fluorescent tubes, to light-emitting diodes (LEDs). These lighting technologies function according to very different physical principles, are available in a variety of form factors, and *prefer* different line voltage-current combinations. Nevertheless, the screw-in socket that Edison invented in 1890 has remained practically unchanged. Without it we would still be stuck with the proprietary wiring and lighting technology solutions that dominated early electricity installations.

FIGURE 4.1 Edison's light bulb: the Sun's only rival. Pictures courtesy of The Smithsonian Institution. (From Carl Sulzberger, A bright and profitable idea: Four decades of Mazda incandescent lamps, *Power & Energy* 4, 3 (2006): 78.)

Here is how Edison described the invention and the value of his solution:

> My object is to adapt my incandescent electric lamps for use not only with the socket ordinarily employed in my systems of electric lighting, but also with sockets employed in other electric-lighting systems.
>
> This interchangeable feature is frequently of great value—for example, when a building has been wired and equipped by one lighting company and for some reason it is desired to substitute my lamps. In this case with the interchangeable terminals the ordinary sockets may be left in place and used in connection with the substituted lamps.
>
> **—Thomas Edison, US Patent 438,310 [26]**

In system terms, the socket works as an interface between the Tool (light bulb) and the Distribution (electric wiring). It enables easy replacement of the light bulb and allows for unimpeded* passage of electricity that powers the bulb. Because the socket sits between two elements with vastly different

* In some instances, however, one may want to control the flow of electricity through the interface more closely; for example, based on ambient light conditions or motion in the room. These modifications, however, do not change the basic form and function of the interface.

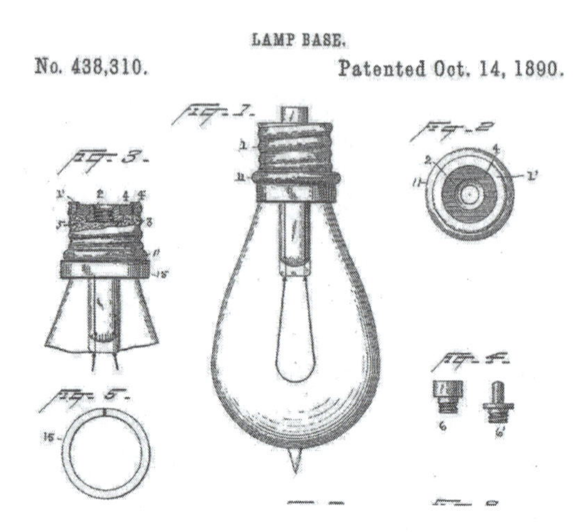

FIGURE 4.2 Edison screw-in socket, US Patent 438,310.

implementation and innovation life cycles, its design managed to outlive both of them.

Before the invention, to install a light bulb one had to hire a trained electrician. It was expensive and inconvenient for many customers. When GE included Edison's standard socket in its technology and brand licensing program, the company created a huge market where various parties could independently invent and perfect their own technologies: incandescent filaments, bulbs, sockets, switches, electric wiring schemes, and other elements of the system. All these innovations led to an incredible growth of electric systems around the world. The humble, yet efficient screw-in socket has survived decades of innovation and is likely to last well into the twenty-first century. Edison's solution to the interface paradox proved to be astonishingly successful.

INTERFACE EXAMPLE 2: THE UNIVERSAL SERIAL BUS INTERFACE

Like Edison's light bulb socket, universal serial bus (USB) connectors are ubiquitous in the modern world. Whether you want to charge your mobile phone, transfer data to a music player, connect a scanner, or attach a camera to your PC, you had better have a USB cable or a built-in USB connector. Today, USB is a key interface between systems that require transfer of information and/or portable electric power.

The USB standard was originally developed in the 1990s to replace a gamut of connectors on the rear panel of any given personal computer (PC). Intel, a major industry force behind the solution, saw the interface as a critical technology that would position the PC as an essential device in the home and the office. The easier it is to add peripherals to the computer, the more applications you can run on the computer, the more reasons for consumers and businesses to buy applications, computers, and peripherals. Even Apple Computers eventually switched to USB, although originally the company adopted FireWire as its own high-speed PC connectivity solution. Recently, electric companies got into the USB game as well. For example, today GE markets power outlets with USB-compliant sockets that can charge your mobile devices without an external power adapter. Leveraging its interface position within the IT world, the USB standard has become a default option for powering portable devices, transferring data, and sending and receiving control signals.

Like any other successful system interface, USB enables a wide range of device and application innovations on both sides of the connector, for example, between traditional laptops and new tablets, which didn't exist in the early 1990s. Today, in the third generation of the technology, mechanical and most electrical properties of the interface are still compatible with those of twenty years ago. For example, you can plug an ancient USB 1.0 mouse into the latest USB 3.0 laptop slot and it will still work. The original USB 1.0 bandwidth of 1.5 Mb/s has been and still is sufficient for sending mouse clicks. But for newer devices, such as tablets and media players, the data transfer rate can be three thousand times higher—up to 4.8 Gb/s. Paradoxically but not accidentally, the USB solution enables stability *and* rapid innovation.

From a business perspective, USB is a good example of an innovation model where intellectual property (IP), including patents and trademarks, has been used from the very beginning to speed up technology adoption. The model requires any company joining the USB standardization effort to license its IP to other USB Forum members on royalty-free terms. Intel and its partners used their patent portfolios and market muscle to include, rather than exclude, third-party developers. This approach reduces patent litigation risks and lowers technology adoption costs.

INTERFACE EXAMPLE 3: APPLE'S 30-PIN CONNECTOR

An alternative interface-related business model is to use a de facto standard for generating licensing revenue for its owner. Similar to GE's strategy discussed earlier and unlike the USB Forum's approach, Apple took the proprietary route when developing a connector for the iPod, iPhone, and iPad suite of devices (Figure 4.3).

The connector allowed new portable gadgets to interface with Macs, PCs, and audio-video systems, independently of the evolution cycle of the input/output ports on the laptop or PC. For example, until relatively recently, to

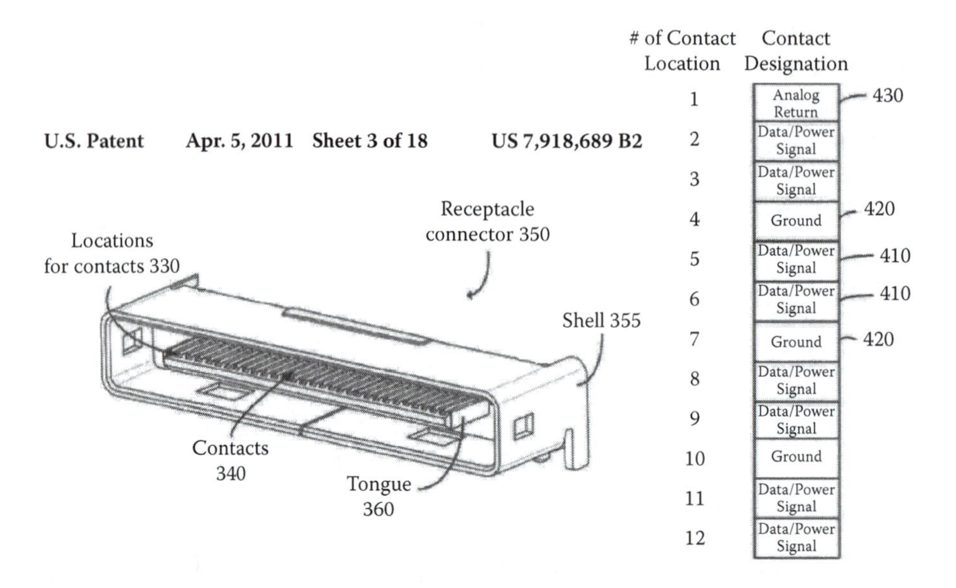

FIGURE 4.3 One of Apple's 30-pin connector patents, US Patent 7,917,689.

transfer your content and apps from a laptop to an iPhone, one had to use a cable with a USB interface on one side and the 30-pin connector on the other. If you wanted to play music through an iPhone-compatible car audio system, you had to plug the device right into the car's 30-pin dock. Since Apple owns a number of patents on the interface, manufacturers of the cable itself and the car audio system must license Apple's IP and pay royalties.

Besides making money for Apple, the proprietary interface allows the company to innovate independently from the standards bodies, who can be rather slow in making their technology decisions. The 30-pin interface on the portable device decouples innovation cycles on both sides of the connection cable. For example, the first iPods came out with FireWire plugs. Then, when Apple decided to connect them to Windows-based PCs, the company introduced USB cables. Now, when DisplayPort and other high-bandwidth slots become available on the PC side, the company has multiple options to adapt the physical characteristics of the interface on its own devices—iPod, iPhone, iPad, and iNext. The control over the interface allows the company to influence the pace of the innovation among system elements and capture hardware license fees from companies that make complementary products.

Although the mechanical and electrical properties of the interfaces on the PC side (such as the USB) are standard, the command-and-control protocol Apple uses to communicate to portable devices, for example, between the iPhone and iTunes PC software, is proprietary. It has become a part of the IP package that Apple licenses to third parties. Moreover, the protocol itself depends largely on the functionality of the device, not the physical features of the connector. For example, should Apple decide to go completely wireless and get rid of the physical dock altogether, the protocol can be implemented to work via the wireless instead of wired connection.

In system terms, the protocol is a part of the Control functionality, which interacts with the rest of the system through software application programming interfaces (APIs). If an unscrupulous cable manufacturer pirates the physical connector, or the connector disappears completely in favor of wireless access to Internet-based services, the protocol and the software APIs will remain a major Control Point for Apple.

Edison invented a great interface with the screw-in socket. But Apple inventors did even better: They solved the interface paradox with not one, but two system components at once—the Distribution (cable and wireless) and the Control (protocols and APIs).

INTERFACE EXAMPLE 4: USER INTERFACES (UIs)

The QWERTY keyboard layout was invented over a hundred years ago (Figure 4.4). Its purpose was to reduce typing speed and prevent the jamming of mechanical parts inside the typewriter. Since then, generations of people learned how to use the immutable interface for interacting with typewriters, teletypes, mainframe computers, personal computers, and smartphones. Mechanical jamming is no longer a problem that can be solved by slowing down one's typing on a virtual keyboard, but the QWERTY interface remains an island of stability in our fast-changing world of digital communications.

Now that we understand the need for interface consistency, a change in an interface should serve as an indicator of a major change within the systems it bridges. For example, the automobile (a horseless carriage), cannot be driven using the same interface as the horse. Thus, we have a steering wheel instead of leather reins. The wheel itself was borrowed from sailing, where one had to exert a significant physical effort to move the rudder. Because early cars did not have power steering, the wheel lever mechanism was adapted to a new application. Today, we still use this centuries-old interface, despite the fact that a vast majority of automobiles have power steering and controlling them is not physically demanding. Furthermore, many modern automobiles employ drive-by-wire technology, which eliminates the mechanical link between the user interface element and the rest of the drivetrain. This shift indicates a likely transition to self-driving cars and the disappearance of the steering wheel altogether.

In another example, the computer mouse, invented by D. C. Engelbart in the 1960s (Figure 4.5), became widely adopted as an essential hardware

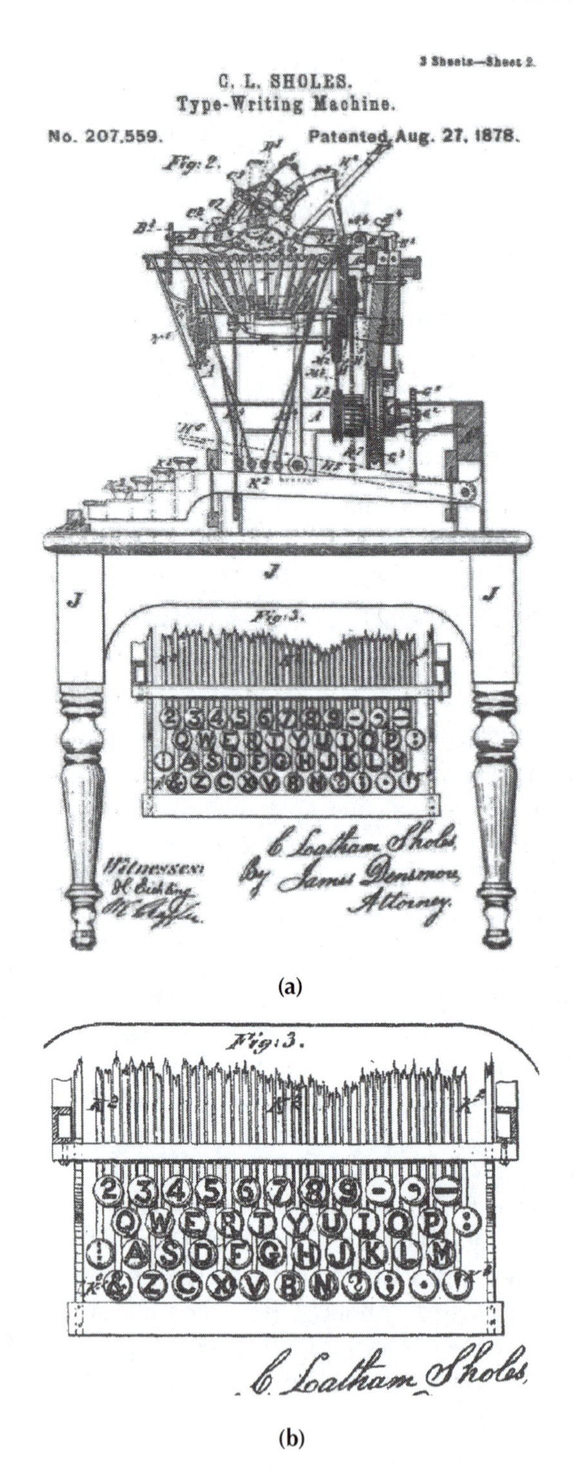

FIGURE 4.4 (a) and (b) C. L. Sholes' typewriter US Patent 207,559.

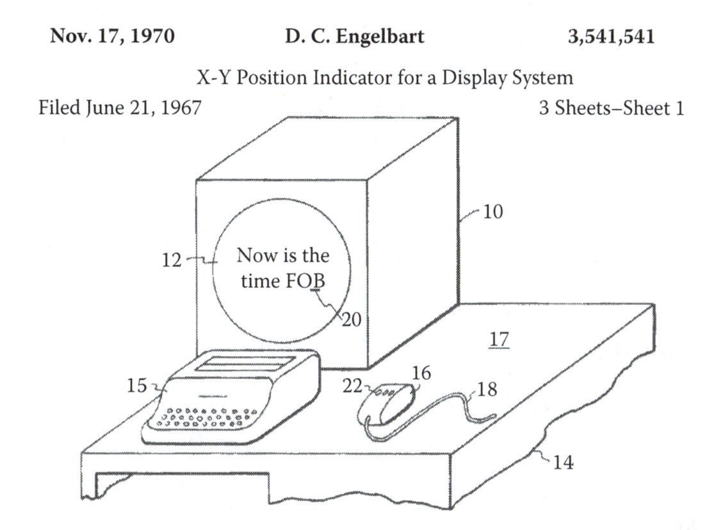

FIGURE 4.5 D. C. Engelbart's computer mouse.

component of the graphical user interface (GUI). The advent of the PC made GUIs ubiquitous. Any PC user could notice that over the last thirty years the world of computing underwent an incredible amount of hardware and software changes. Technical characteristics of a typical consumer PC (e.g., processor speed, storage space, and memory) increased at least a thousandfold each. On the other hand, the GUI did not change much. On all mainstream PCs, either Apple Macs or Microsoft Windows, you find icons and document windows you are supposed to interact with using the mouse. Even the World Wide Web, with the notable exception of the hyperlink, uses the same UI paradigm as the PC. The astonishing rate of innovation in hardware and software seemed to bypass the GUI. This is no longer the case, however. The disappearance of the mouse on tablet computers and smartphones suggests to us that a major system-level transformation is under way. Is it going to be as dramatic as the transition from the horse to the automobile? Yes, would be the most likely answer.

In Section III of this book, we will use system analysis to examine the extent and possible implications of the transformation. In the meantime, it's important to note the significance of recent practical, widely adopted innovations in the area of human–computer interfaces, such as Apple's multitouch screen,* voice control with Siri, and Microsoft Kinect. Old UIs die hard. The latest changes hint at the emergence of fundamentally new capabilities in the world of computing and communications.

* While Apple Inc. was not the originator of the concept and demonstration, it was the first entity to realize its wide commercial use.

As inventors or innovators we should treat such indications of system-level change as once-in-a-lifetime opportunities. Though Apple, Microsoft, and other parties patented heavily around the new interfaces, some other critically important elements of the system remain open. Since we cannot invent everything at once, we need to learn how to focus our efforts and resources on problems that are worth solving. In the next chapter we will discuss *Control Points*—a system-based technique for discovering high-value problems.

REVIEW QUESTIONS

1. In April 2012, Google began testing its augmented-reality glasses. According to the *New York Times*, "The prototype version Google showed off on Wednesday looked like a very polished and well-designed pair of wrap-around glasses with a clear display that sits above the eye" [27]. Outline a system where the glasses play the interface role. What is the primary type of the Packaged Payload that flows through the new interface: mass, energy, or information? What interface devices can be replaced by the glasses? How would this change affect the pace of innovation? Explain and give specific examples.
2. In recent years, a number of companies introduced novel electronic interface devices intended to replace traditional electric and mechanical interfaces, such as door locks [28] and heat, ventilation, and air conditioning (HVAC) thermostats [29]. Considering the implementation of system element(s) (i.e., the Tool, Source, Distribution, Packaged Payload, or Control), which specific system elements are likely to grow in importance due to the interface innovations? Why?
3. Imagine you are walking a dog on a busy city street.
   a. What interface(s) would you use to interact with the dog?
   b. Would you use a different interface if you were to run with the dog on a remote ocean beach? Explain.
   c. Would you use a different interface if you were to run with a pack of 100 dogs? Explain.
   d. Would you use a different interface if the dog's weight were 10 pounds, 60 pounds, or 160 pounds? Explain.
   e. What interface would you use to interact with an imaginary dog? Discuss with the dog and explain.

5 System Control Points
Where to Aim the Silver Bullets

During WWII, statistician Abraham Wald was asked to help the British decide where to add armor to their bombers. After analyzing the records, he recommended adding more armor to the places where there was no damage! [30]

The expert's advice was based on the insight that airplanes with damage to important parts probably never returned to their airbases. On the other hand, only those with relatively inconsequential types of damage were the ones that managed to return. Contrary to the initial intuition, the armor—that is, *control* over potential damage—had to be applied to parts that stayed intact on the inspected airplanes.

Using a similar approach, we look for Control Points by finding the most reliable ways to break a system. Once they are found, we target them as key problem tasks for further invention work.

There are many simple ways to break a system temporarily. One way would be to destroy an instance of a specific system element. For example, earlier in the book we discussed various inconveniences our friend Bob encountered after he lost his iPhone. A similarly annoying problem would be finding the car battery dead in the morning before going to work. The same is true for a missing laptop power supply or a burnt-out light bulb in the bathroom. These problems, although unpleasant in the short term, can be easily solved by widely available substitutes. You can call for a taxi, order or borrow a power supply, replace the light bulb, or use a flashlight or a candle in the meantime. The more mature the system, the more substitution options are available for you to choose from. To inflict permanent damage on the system we have to be more sophisticated.

Particularly, we want to find situations when all elements by themselves are in perfect order, but the system as a whole remains broken. In such cases, a substitution would not solve the problem. For example, consider a situation when you need to send an urgent confidential message. You've got your smartphone in your hand, it's in perfect working order, and has all the relevant contact info. However, the device is password protected and you have forgotten the PIN. The system doesn't work and you can't use somebody else's phone either.

Another common frustration is when you have a Mac laptop and need to make an important presentation at a company that has a policy permitting only Windows-based PCs on the premises. To connect to an overhead projector, you need a special adaptor, which you happen to forget at the place of the previous meeting. If your wish is to show the presentation, without the adaptor, the sophisticated computing environment around you is no more useful than the furniture in the room.

In one more example, imagine yourself surrounded by skyscrapers in a busy downtown of a foreign city. The global positioning system (GPS) unit in your rental car is in perfect working order and it has the destination programmed in. However, there's no reception. That is, all the machinery is in place but if you are at an intersection, you don't know where to go. Like a World War II bomber airplane, you (or the system you are relying on) got hit in a highly sensitive spot, and it's practically impossible to accomplish the mission.

When we map these problem situations on the system model we discover that systems are most fragile at interfaces between components, Packaged Payloads, and often the Control subsystems. Figure 5.1 shows the weak spots one can attack to destroy even a sophisticated implementation. The lost PIN prevents you from accessing the interface of a perfectly intact smartphone. The missing connector does not let you hook up hardware interfaces between the laptop and the projector wiring. The lost GPS signals prevent the delivery of info payloads necessary for your GPS navigator to make routing decisions.

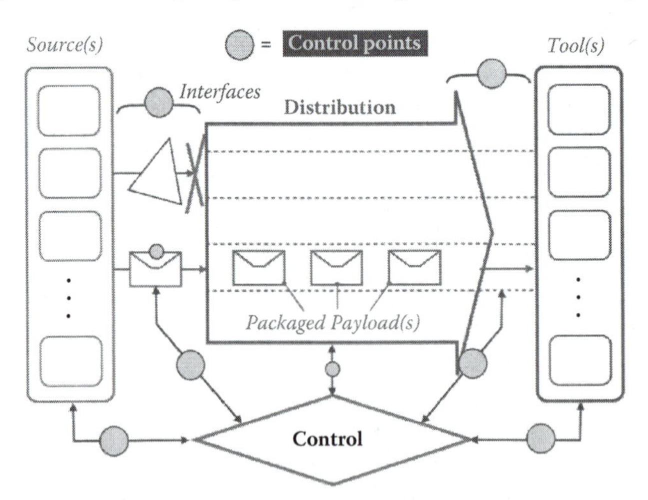

FIGURE 5.1 A system diagram with Control Points and interfaces.

The missing iTunes software prevents Bob from enjoying his new songs and apps. The system doesn't work despite the fact that all of its other components remain intact.

A patent's power to exclude others is best deployed on Control Points, enabling a legal attack on the fragile aspects of any viable implementation.

CONTROL POINTS AS OPPORTUNITIES FOR INNOVATION

Once a Control Point is compromised, the entire system may become completely irrelevant. The War of Currents between Thomas Edison and George Westinghouse is a good example of such conflict.

By the mid-1880s, Thomas Edison developed an end-to-end electric grid solution based on low-level direct current (DC). His power stations, parallel wiring schemes, metering equipment, and light bulbs worked really well together, bringing lights to streets, offices, department stores, and houses. Due to the combination of the inventor's ingenious technical solutions, strong patents, and public relations savvy, competing with him in the DC market was extremely difficult for electricity entrepreneurs.

While the DC-based system was a great improvement over the previous generation of technology, its applications were limited to about a one-mile radius from the power station. This constraint would not let it scale up to accommodate the growth of American industry and commerce toward the end of the nineteenth century. For example, if you wanted to power a large section of the city or a couple of big factories, you had to buy and build multiple power stations and ensure large-volume, steady supplies of coal to feed the stations' steam engines. The model worked well for Edison's manufacturing businesses and railroad owners, but was not efficient enough for broad market adoption.

> I never think of the past. I go to sleep thinking of what I am going to do tomorrow.
>
> **—George Westinghouse [3, p. 337]**

George Westinghouse (1846–1914), an American inventor and entrepreneur, made his early fortune in railroad technology. His invention of the steam-power brake for train cars (US Patent 88,929) made him a millionaire at the age of 23. In the 1880s, Westinghouse recognized the great business potential of electric power. Instead of competing with Edison head to head in the DC market, Westinghouse placed his bet on the alternating current (AC)

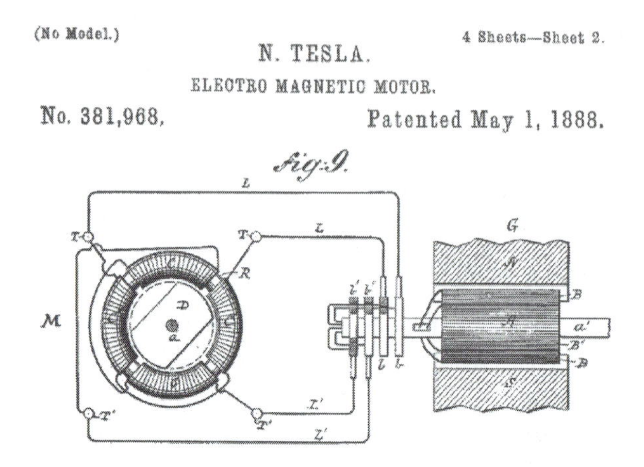

FIGURE 5.2 One of Nikola Tesla's electric motors, US Patent 381,968.

approach. Unlike DC, AC could be generated on a large scale, transferred via high-voltage power lines over long distances, and converted to low voltage using transformers. The technology approach looked promising, but to make it a business success, Westinghouse had to invest massively in practical implementations for all major system elements.

And he did it. The world we live in today runs on an AC electric power infrastructure. In contrast to Edison, Westinghouse staked his innovation strategy on buying or licensing inventions from other people, rather than inventing everything himself. Today, we would call this approach open innovation. One of the most important of his acquisitions was polyphase AC motor patents from Nikola Tesla (Figure 5.2). Tesla's design proved to be critical for the success of industrial applications of electricity around the world. Nowadays, most electricity is used in the form of 50- or 60-Hz AC because over a hundred years ago that was the optimal frequency for Tesla's motor. By working closely with many inventors, both in the United States and Europe, Westinghouse quickly caught up with Edison's invention factory.

Under threat from AC technology, Edison started the War of Currents, a vicious public relations campaign portraying AC as highly dangerous to humans and animals. In forceful attempts to prove his point, he publicly electrocuted dogs, cats, and even an elephant. Death by electric chair is one of the legacies of that time. An AC-powered chair was considered to be a modern, and somewhat humane way to die for people who committed capital offenses. Eventually, Edison lost the War of Currents, because the advantages of AC over DC had become obvious, especially in large-scale deployments. For example, Henry Ford's mass production operations were powered by AC motors. The light bulb remained one of the few practical devices used in both AC and DC power networks. As a result, for ordinary consumers the monumental change in technology went largely unnoticed.

But when we look from a system perspective, the significance of the transition becomes apparent. As long as Edison controlled the Packaged Payload (DC electricity), he controlled the evolution of the system as a whole. Once DC lost to AC, most of Edison's solutions, however ingenious they were originally, became irrelevant. Because the Packaged Payload interacts with all system elements, its change leads to systemwide changes.

CONTROL POINT EXAMPLE 1: MICROSOFT OFFICE VERSUS GOOGLE DOCS

Consider our friend Bob from the previous chapter when he creates and edits documents using Microsoft Office software. For example, when he wants to develop a presentation for his client, he opens up a new file in Microsoft PowerPoint, puts together his slides, and saves the file in the Documents folder. When he wants to share the presentation with his colleague Alice, he emails the file as an attachment, which can be opened with PowerPoint on Alice's computer. The software recognizes the file as a PowerPoint presentation by the file's extension. For example, the latest version of PowerPoint uses *pptx* as the last four letters after the dot in the name of the file.

By now, you've recognized that in system terms the file represents an instance of the Packaged Payload. To work with it, either on the Source or Tool side, Bob and Alice have to have an application that knows how to interpret the file's formatting information. Another piece of software—usually a file browser playing the role of the Control—has to recognize the file and invoke the right application for editing it. For example, if Alice opens it with the wrong app she will see strange characters instead of Bob's slides.

Conveniently for Bob and Alice, Microsoft sells software for all the system elements. That is, the Microsoft operating system enables users to store, recognize files, and invoke appropriate applications. Microsoft email software helps distribute the files. Microsoft applications let users create and edit the files. As you can see, Microsoft practically owns the entire system, which lets the company reap handsome profits in the world of personal computers.

If we were to attack the system, where should we aim?

In Google We Trust: Leveraging Control Points

In 2006, Google acquired Upstartle, a four-person startup that developed Writely, a web-based word processor, which enabled collaborative document editing. If Bob were to use Writely instead of Microsoft Word, he would create and edit his document in a web browser pointed to an Upstartle server. To share the document with Alice, Bob would send her a link rather than the document itself. In the Upstartle world, the file seemed to disappear, masking from Bob and Alice the details of how and where their content was being stored and managed. Although editing and formatting capabilities offered by Upstartle were rather humble, the model proposed by the startup featured a new type of the Packaged Payload, which hinted at a potential collapse of Microsoft's file-based system as a whole.

Let us consider another Google technology acquisition, so that we gain additional insights into the nature and implications of the system-level change taking place in the context of online, real-time document creation and collaborative editing. In 2005, the Internet search giant bought 2Web Technologies, a startup that was developing a web-based spreadsheet product called XL2Web. The concept behind XL2Web was very similar to that of Writely. That is, instead of creating, editing, and sharing spreadsheets using specialized desktop applications, users like Bob and Alice could perform all these tasks in a general-purpose web browser. They could also upload their existing Microsoft Excel spreadsheets to the server and convert them into a web-compatible format. Since spreadsheets, with their formulas and frequent value recalculations, are more dynamic than a typical text document or a presentation, XL2Web implementation had to rely heavily on scripts embedded in the web page. When downloaded from the server, the script would be executed as a lightweight software application inside the web browser. In system terms, within a web-based document system, the script had become the dominant form of the Packaged Payload.

The user experience with both XL2Web and Writely depended to a large degree on the web browser's ability to execute scripts in a fast and reliable manner. Because at the time Microsoft had the lion's share of the web browser market, the company was in a good position to control the pace of innovation in the web-based document system. Once Google acquired XL2Web and Writely, making them parts of the Google Docs suite, the need for Google to build its own browser became more urgent: without a powerful script execution engine inside the browser, Google Docs as well as the rest of Google Apps would have to depend on Microsoft in evolving its product development strategy.

> Page and Brin wanted Chrome optimized to run web applications—fast. When you run a program faster by an order of magnitude, you haven't made something better—you've made something new. [31, p. 208]

Although Eric Schmidt, then Google's CEO, was reportedly against building their own browser, Google funders Larry Page and Sergey Brin authorized building it anyway [32]. By 2008, the company had all system elements in place to challenge Microsoft Office.

Let us identify each element explicitly. On the Source side, Google has powerful Internet servers capable of storing and retrieving millions of documents per second. When Bob and Alice access the servers, they get to use tiny slices of the functionality of the Source.

On the Tool side, Google offers their Chrome browser optimized for JavaScript execution to speed up Google's various online apps. Internet and web protocols play the role of the Distribution element.

Dynamic web pages with embedded scripts serve as instances of the Packaged Payload. The scripts enable interactive features for document creation, editing, sharing, and tracking. Essentially, the notion of any given document as a discrete file has disappeared altogether.

Furthermore, while executing their mission to "organize the world's information," Google has gained significant control over how user content is stored, accessed, shared, secured, and tracked. That is, once Bob and Alice buy into the system, they can't get out of it without serious loss of time and functionality. The more content they develop, the wider they share it, the more of the Control functionality they delegate to the system that is set up and maintained by Google. In short, once they decide to give up on treating their documents as files, the collapse of Microsoft's file-based Office system becomes imminent.

Of course, this doesn't mean Microsoft is going to give up without a fight. As of this writing, the company still has a large share of the desktop web browser market.* Internet Explorer is also dominant in enterprise office software, and with proper technology, could help its customers move toward the new model. Furthermore, in the growing segment of smartphones and tablets, the web is one of *many* connected applications. Therefore, the battle between Microsoft and Google for ownership of the Packaged Payload is not over yet. No matter who wins the battle of the "offices," they are going to provide products and services within a different system.

* "Google's Chrome is now the most popular Web browser worldwide, surpassing Microsoft's Internet Explorer for the first time, according to the latest figures from StatCounter. After years of slowly chipping away Internet Explorer's market share, Chrome took the lead with 32.76 percent share, while IE dipped to 31.94 percent" (http://www.pcworld.com/article/255886/google_chrome_overtakes_internet_explorer.html).

CONTROL POINT EXAMPLE 2: PRIVACY IS DEAD. LONG LIVE E-COMMERCE!

According to the research conducted by Aleecia McDonald and Lorrie Faith Cranor at Carnegie Mellon University, if you were an average Internet user, it would take you 76 normal (8-hour) working days, or almost 4 working months to read all of the privacy policies you had agreed to in the course of your online activities [33]. Unless you are a privacy lawyer being paid for this kind of work, it would be a safe bet that you didn't thoroughly read the policies to which you agreed.

On March 1, 2012, Google implemented a new privacy policy [34] that allowed the company to track individual users across 70+ Google web properties, including search, Gmail, YouTube, Google Maps, and others. Once you have agreed to the policy, which many of us did without reading, Google obtained your permission to collect personal information about your Internet access devices, online activities, content and services, locations, and apps. Moreover, you agreed to Google sharing aggregated forms of this information with its advertisement partners.

For the purposes of our Control Points discussion, it is important to note how Google implemented user tracking. Here's what the policy says:

> We use various technologies to collect and store information when you visit a Google service, and this may include sending one or more cookies or anonymous identifiers to your device. We also use cookies and anonymous identifiers when you interact with services we offer to our partners, such as advertising services or Google features that may appear on other sites.
>
> **—Google Privacy Policy, 2012**

In short, when you get a web page with the content you've requested from a Google service, you also get one or more cookies—small files containing identification information. Google and their partners' services can read and modify the cookies, so that the information about what, when, where, why,

> Google Inc. agreed to pay $22.5 million, the largest fine ever levied by the U.S. Federal Trade Commission, to settle allegations that it breached Apple's Safari Internet browser.
>
> The record fine is the FTC's first for a violation of Internet privacy as the agency steps up enforcement of consumers' online rights.
>
> **—*Bloomberg News*, August 9, 2012 [35]**

and how you use the web can be collected and analyzed. If you set your browser to reject the cookies, your personalized services, such as Gmail, Google+, Google Docs, and others will not work. To summarize, *no cookies means no service.*

From the system model perspective, the cookie represents the *aboutness* of the interaction between different system elements, e.g., between instances of the Source and the Tool. The aboutness enables the system to collect information and use it to orchestrate the performance of its elements or of the system as a whole. Furthermore, it facilitates the building of processes comprising a daisy chain of various instances of the Source and the Tool. The aboutness is often embedded into, sent along with, or ahead of the Packaged Payload.* Through the use of the aboutness the system lets the Control element orchestrate a process involving multiple interactions between system elements.

For example, in the Google Docs system the cookie contains web browsing session information, so that you don't have to log into the Google Drive server every time you access a different document. In another example, Google Analytics can use cookie information and single-pixel tracking images to accumulate statistical data from multiple users to see how they respond to a particular advertisement campaign. As a company, Google thrives on large amounts of user data because it allows the advertiser (or Google) to customize individual web pages, and to fine-tune server and browser performance. The vast majority of web services, including Google, Facebook, Amazon, eBay, Twitter, and other popular sites would not work if cookies are disabled in the browser. It would not be an exaggeration to say that without the cookie concept, or some other implementation of the aboutness, e-commerce or social networking on the web would be impossible.

The World Wide Web (as proposed by Tim Berners-Lee) was originally designed as a technology for sharing document *files* between scientists. Because the scientists worked at different research institutions, they used disparate computer systems and couldn't access and reference each other's work in an easy, transparent way. The web provided them with a simple technology to perform a single download transaction between multiple computers. As the Packaged Payload, the *web page* was born in the 1980s. It took the world another decade to invent an implementation of its aboutness—the cookie.

The honor of enabling the web as we know it today belongs to Lou Montulli, a founding engineer at Netscape Communications. In 1994, he applied to the

* Numbers and watermarks on a banknote are other examples of the aboutness. The numbers show how much currency value the banknote represents; the watermarks prove that the banknote is legitimate.

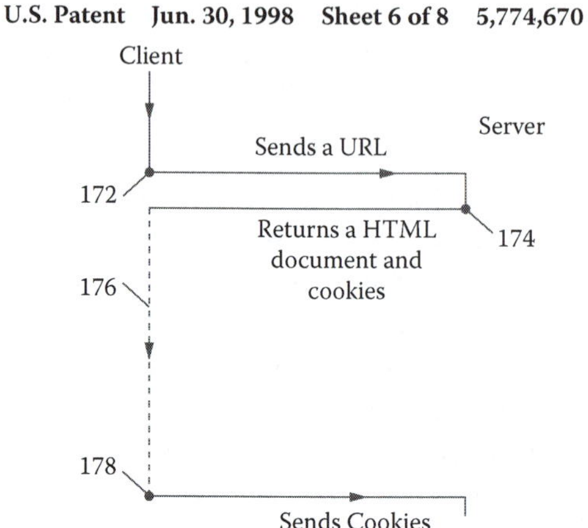

FIGURE 5.3 The web cookie, US Patent 5,774,670. The browser (client) sends a web page address (URL) to the server. The server returns the page with cookies. The browser sends the cookies when requested.

Netscape Navigator browser the concept known among programmers as a *magic cookie* (a token of data exchanged between communicating programs). In 1998, he was awarded a patent (US patent number 5,774,670) for "Persistent client state in a hypertext transfer protocol based client-server system" (Figures 5.3, 5.4). Though Montulli's name is not as widely recognized as that of Tim Berners-Lee, he got his share of fame and fortune. In 1999, he was named the *People* magazine sexiest Internet mogul of the year. In 2002, *MIT Technology Review* named him among the top 100 innovators under 35.

From the system perspective, it is no accident that after Montulli's invention the use of World Wide Web among nonscientists took off. Having the ability to implement and use the cookie helped entrepreneurs bring e-commerce and other web services to the larger public. When we analyze or invent systems, it is important to remember the aboutness aspect of the Packaged Payload. As we can see from the Google and other examples, knowing what's going on with system elements is critical for effective, scalable implementation of the Control functionality.

CONTROL POINT EXAMPLE 3: THE APPLICATION PROGRAMMING INTERFACE

In previous chapters, using examples of the Edison socket and Apple's 30-pin connector, we discussed the importance of interfaces for enabling rapid, independent innovation among different system elements. In information systems, a set of application programming interfaces (APIs) plays the role of

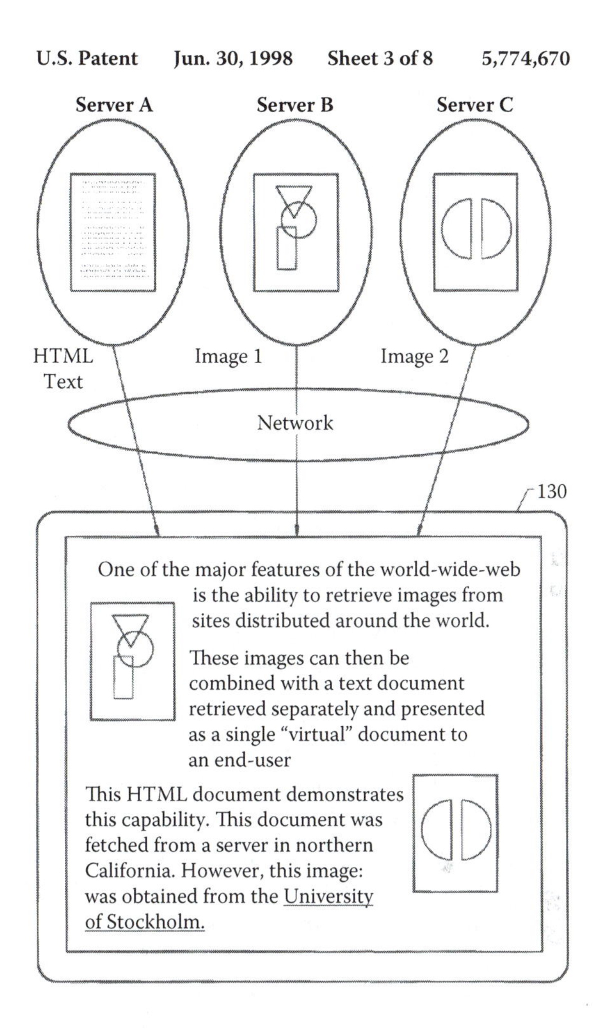

FIGURE 5.4 A web page diagram from US Patent 5,774,670.

the interface. For example, when a user edits his Google Drive document, the script running inside his browser uses certain APIs to talk to the server. Just like the interfaces we discussed earlier in the book, the software interface allows for a different pace of innovation on either side of that interface.

Google Drive is an integrated, relatively secure system and it does not give developers a lot of free access to user-generated content. By contrast, Facebook represents an example of an Internet-based system designed with the explicit goal to provide APIs for sharing user data with third-party applications. In May 2007, just months after it opened up Facebook membership to the general public, the company launched Facebook Platform, which enabled developers to create apps that could access information hosted by Facebook. In December 2008, Facebook released Facebook Connect APIs, which let members use their social network identity to log onto other websites and devices. In early 2012, Facebook introduced APIs that expose user timelines to third-party developers. By now, any web page, application, device, or connected thing can embed software that allows it to feed information into and receive information from Facebook.

As you can see, the APIs make it easy for third parties to add elements to the Facebook system, powering its rapid growth. Think of the system as a worldwide attention exchange. If there's something worth paying attention to among a billion users of Facebook, the company is in a position to offer this attention-worthy content to the broader attention market. Furthermore, because the information accumulates over a long period of time, it becomes possible to understand, predict, and influence user interests and attention patterns. Once you've got someone's attention, you can present him or her with an ad or make a sale. Among all the players in the system, Facebook is the only party that has access to the totality of social information. By sharing slices of the information with third parties through the APIs, Facebook creates opportunities for discovering and experimenting with novel ways of generating and sharing user interests, from gaming to shopping to making revolutions.

The users and customers of the service have delegated control over their data to Facebook: how it is collected, processed, organized, packaged, forwarded, and received. If any of the APIs that facilitate these operations break down, the system would collapse. That is, hundreds of thousands of apps would fail, millions of people would lose access to their social lives, and billions of web pages would have to be redesigned through no fault of their own. Because Facebook owns the APIs and their implementation, the company has in its firm possession the vital Control Point of the system as a whole.

REVIEW QUESTIONS

1. Give examples of the aboutness routinely found in most supermarkets. Who uses the various types of aboutness you have identified, and why?
2. You are the CEO of the Stem startup company discussed in an earlier chapter. What is your potential Control Point?
3. In 2012, Apple introduced the iPod 5 smartphone with the lightning connector that replaces the old 30-pin connector. According to cnet.com, "Apple sells a $39 two-meter-long Lightning-to-30-pin cable, and a regular Lightning-to-USB cable costs $19" [36]. Reportedly, the connector has an embedded authentication chip intended to

prevent use of counterfeit, unlicensed third-party cables and to enable other functions. What type of the Control Point does the new connector implement? Will the connector increase, decrease, or leave unchanged the pace of innovation in the Apple iPhone system? Explain your choice.

4. In 2012, German researchers discovered what seems to be room-temperature superconductivity in graphene, or more specifically, in the boundaries between grains in graphene. While the effect is extremely small, let's imagine that somehow it could be enhanced, and the superconducting material could be produced in large quantities. Introduce a superconductor into some system of interest. Systematically analyze what you could do to or with each of the system components that you couldn't before? (Be sure you systematically evaluate in instances where the Packaged Payload is mass, energy, information.) What would happen to the entire system in each instance? How would these transformations improve the functionality of the system from the standpoint of the user? What are the key interfaces and Control Points?

CONCLUSIONS

In Section I we introduced the system model and its basic elements: the Tool, Source, Distribution, Packaged Payload, and Control. We showed how the model relates to real-world technologies and patents. Initially, we focused on the elements' interactions inside the system: how the system expands, interfaces develop, and Control Points emerge. Although we introduced some recursive features of the model, we mostly used them to zoom inside the system "box." In Section II, we are going to learn how to look outside the box and navigate freely between multiple system levels. Our main mission will be to discover what makes some inventions luckier than others.

REFERENCES FOR SECTION I

1. Silverstein, Shel, *Where the Sidewalk Ends* (New York: Harper and Row, 1974).
2. Altshuller, G.S. Tvorchestvo kak tochnaya nauka, Moskva (1979). http://vikent.ru/enc/386/
3. Jill, Jonnes, *Empires of Light: Edison, Tesla, Westinghouse, and the Race to Electrify the World* (New York: Random House, 2004).
4. Whitney, Lance, Edison tops Jobs as world's greatest innovator, c|net, January 26, 2012, http://news.cnet.com/8301-11386_3-57366904-76/edison-tops-jobs-as-worlds-greatest-innovator/

5. Edison, T.A., Webermeter. US Patent 240678, filed Oct 7, 1880 and issued Apr 26, 1881.
6. USPTO Press Release, November 14, 2011, http://www.uspto.gov/news/pr/2011/11-67.jsp
7. Helft, Miguel, and Shan Carter, A chief executive's attention to detail, noted in 313 patents, *New York Times*, August 25, 2011, http://www.nytimes.com/2011/08/26/technology/apple-patents-show-steve-jobss-attention-to-design.html
8. Smil, Vaclav, Why Jobs is no Edison, *The American*, September 30, 2011, http://www.american.com/archive/2011/september/why-jobs-is-no-edison
9. Kahneman, Daniel, *Thinking Fast and Slow* (New York: Farrar, Straus and Giroux, 2011).
10. Nokia's bad call on smartphones, *The Wall Street Journal*, July 18, 2012. http://onlinewsj.com/article/SB10001424052702304388004577531002591315494.html
11. iMovies this month? Sept 1, 2006. http://news.cnet.com/8301-10784_3-6111730-7.html
12. Reardon, Marguerite, AT&T's CTO defends wireless network, c|net, October 8, 2009, http://reviews.cnet.com/8301-12261_7-10371298-10356022.html
13. Townsend, Matt, Best Buy sales at risk as surgical shoppers lose impulse: Retail, *Bloomberg Businessweek*, October 17, 2011, http://www.businessweek.com/news/2011-10-17/best-buy-sales-at-risk-as-surgical-shoppers-lose-impulse-retail.html
14. Reisinger, Don, iOS web traffic surpasses Mac OS X for first time? Maybe, c|net, February 10, 2012, http://news.cnet.com/8301-13506_3-57374686-17/ios-web-traffic-surpasses-mac-os-x-for-first-time-maybe/
15. Facebook mobile users exceed desktop users. Jan 30, 2013. http://www.marketwatch.com/story/facebook-mobile-users-exceed-desktop-users-2013-01-30
16. Topol, Eric, *The Creative Destruction of Medicine* (New York: Basic Books, 2012).
17. Smith, P.K., and Y. Trope, You focus on the forest when you're in charge of the trees: Power, *Journal of Personality and Social Psychology* 90, 4 (2006): 578–596. DOI: 10.1037/0022-3514.90.4.578.
18. Troianowski, Anton, AT&T move signals end of the copper-wire era, *Wall Street Journal*, November 7, 2012, http://online.wsj.com/article/SB10001424127887324439804578104820999974556.html.
19. LaMonica, Martin, A startup's smart batteries reduce buildings' electric bills, *MIT Technology Review*, November 6, 2012, http://www.technologyreview.com/news/506776/a-startups-smart-batteries-reduce-buildings-electric-bills/.
20. Eisenberg, Anne, Their mission: To build a better toilet, *New York Times*, August 13, 2012, http://www.nytimes.com/2011/08/14/business/toilet-technology-rethought-in-a-gates-foundation-contest.html?_r=0.
21. Hyde, R.A. et al. Plasmon Gate. US Patent 7,542,633, filed Oct 12, 2007 and issued Jun 2, 2009.

22. Wozniak, Steve, *iWoz: Computer Geek to Cult Icon* (New York: W. W. Norton, 2006).

23. Sanger, David E., Obama order sped up wave of cyberattacks against Iran, *New York Times*, June 1, 2012, http://www.nytimes.com/2012/06/01/world/middleeast/obama-ordered-wave-of-cyberattacks-against-iran.html

24. Shteyn, Y. Eugene, Visual Anti-Virus in a Network Control Environment. US Patent 7,092,861, filed Nov 2, 2000 and issued Aug 15, 2006.

25. Lemley, Mark A., and Carl Shapiro, Probabalistic patents, *Journal of Economic Perspectives* 19, 2 (2005): 7–98.

26. Edison, Thomas A. LampBase. US Patent 438,310, filed May 5, 1890 and issued Oct 14, 1890.

27. Bilton, Nick, Google begins testing its augmented-reality glasses, *New York Times*, April 4, 2012, http://bits.blogs.nytimes.com/2012/04/04/google-begins-testing-its-augmented-reality-glasses/.

28. Golson, Jordan, Lockitron announces new keyless deadbolt entry and remote locking for iPhone, *MacRumors*, October 2, 2012, http://www.macrumors.com/2012/10/02/lockitron-announces-new-keyless-deadbolt-entry-and-remote-locking-for-iphone/.

29. O'Connell, Frank, Inside the nest learning thermostat, *New York Times*, October 3, 2012, http://www.nytimes.com/interactive/2012/10/04/business/inside-the-nest-learning-thermostat.html

30. Cook, John D., Selection bias and bombers, *The Endeavor: The Blog of John D. Cook*. January 21, 2008. http://www.johndcook.com/blog/2008/01/21/selection-bias-and-bombers/

31. Levy, Steven, *In the Plex: How Google Thinks, Works, and Shapes Our Lives* (New York: Simon & Shuster, 2011).

32. Angwin, Julia, Sun Valley: Schmidt didn't want to build chrome initially, he says, *WSJ* Blog, July 9, 2009, http://blogs.wsj.com/digits/2009/07/09/sun-valley-schmidt-didnt-want-to-build-chrome-initially-he-says/

33. McDonald, Aleecia M., and Lorrie Faith Cranor, The cost of reading privacy policies, 2008, *I/S: A Journal of Law & Policy for the Information Society* 4, 3 (2008), http://moritzlaw.osu.edu/students/groups/is/files/2012/02/Cranor_Formatted_Final.pdf

34. Google Privacy Policy Comparison, Google, http://www.google.com/policies/privacy/archive/20111020-20120301/

35. Forden, Sara. Google said to face fine by U.S. over Apple Safari breach. *Bloomberg News*, August 9, 2012. http://www.bloomberg.com/news/2012-08-09/google-said-to-face-fine-by-u-s-over-apple-safari-breach.html

36. Apple iPhone 5 gives the world a new connector: Lightning. Cnet.com, September 12, 2012. http://news.cnet.com/8301-13579_57511581-37/apple-iphone-5-gives-the-world-a-new-connector-lightning/

Section II

Outside the Box

Right here and now, as an old friend used to say, we are in the fluid present, where clear-sightedness never guarantees perfect vision.

— **Stephen King and Peter Straub,** ***Black House***

6 Outside the Box
Developing Skills for Creative Thinking

To encourage creativity, people often say, "Think outside the box." The recommendation urges divergent thinking, a break from common rules and assumptions. Murray Gell-Mann, a Nobel laureate in physics, believes that the phrase originates from the popular nine-dots puzzle, where the solver has to cross all the dots with no more than four straight lines without lifting the pen [1] (Figure 6.1). The puzzle is impossible to solve if the lines stay inside the box (outlined by the dashed square in Figure 6.1). But once you go beyond the box, the solution becomes almost trivial.

A recent study showed that people perform better on standard creativity tests when they sit outside of a fixed workspace [2]. Maybe simply getting away—even just a little bit—from the familiar walls helps solve puzzles requiring a certain kind of thinking. Other psychological experiments show that a simple mention of a city associated with creativity, such as London or New York, improves one's ability to be more creative on an unrelated task [3].

Of course, the trouble with the advice to "think outside the box" is that most of the time when we deal with difficult problems we don't know where the box is. Unlike puzzles, most problems we try to solve as inventors and innovators don't have predefined answers [4]. Moreover, in many situations finding the right problem is more difficult (and more important) than solving it. Sometimes, when people try to think outside of one box they immediately jump into another box, with no guarantee that thinking inside the new box is more creative than in the old one.

In this chapter we introduce tools for seeing the box that we might be in and the constraints it imposes on our thinking. Then we learn how to create multiple perspectives, get outside the box, and discover high-quality problems that require creative solutions.

In a CNN interview, author and risk management specialist Nicolas Taleb, gives a classical parable of a Thanksgiving turkey destined to die due to a limited time perspective [5]:

To solve the puzzle, connect all 9 dots using four straight lines or less, without lifting the pen.

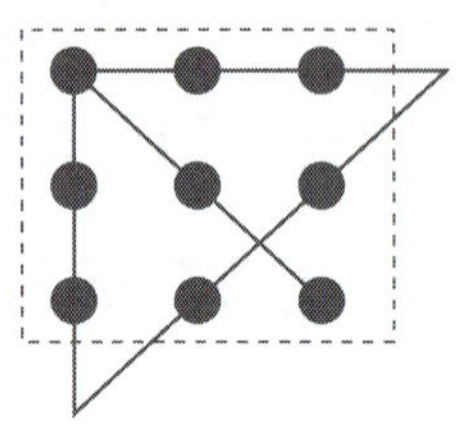

FIGURE 6.1 The 9-dot puzzle.

> A turkey is fed for 1,000 days—every day lulling it more and more into the feeling that the human feeders are acting in its best interest. Except that on the 1,001st day, the butcher shows up and there is a surprise. The surprise is for the turkey, not the butcher.*

Sometimes, such bad surprises are due to bad information deeply embedded in routine processes. For example, when Kenneth Arrow, a future Nobel laureate in economics, served in the army he discovered that his superiors relied on outdated weather forecasts. Arrow's warning that the forecasts were useless got rebuffed:

> The Commanding General is well aware the forecasts are no good. However, he needs them for planning purposes. [6, p. 47]

In the Internet age, though, we can easily discover up-to-date information. But as the turkey parable shows, even when forecasts are based on the right information, the consequences can still be disastrous. Taleb's reference to the turkey alludes to a philosophical discussion about the general problem of induction [7]. (Induction is a method of reasoning when predictions about future events are derived from present and past experiences. Taleb's turkey thinks about the future by extrapolating from his prior experiences.) According to a highly influential skeptical view expressed by philosopher David Hume in the seventeenth century, we simply cannot predict the future by induction [8].

Nevertheless, as inventors and innovators we often must go into the future having a limited time horizon. One way to deal with any turkey-like situation would be to give up and say that it is impossible to foresee and forestall a large-scale problem. Another, more optimistic way of looking at the

* Similarly, in the Prologue we discussed companies that for a period of time made perfectly reasonable trade-offs, only to find themselves in the "surprised turkey" situation in which a competitor eliminated the need for trade-offs.

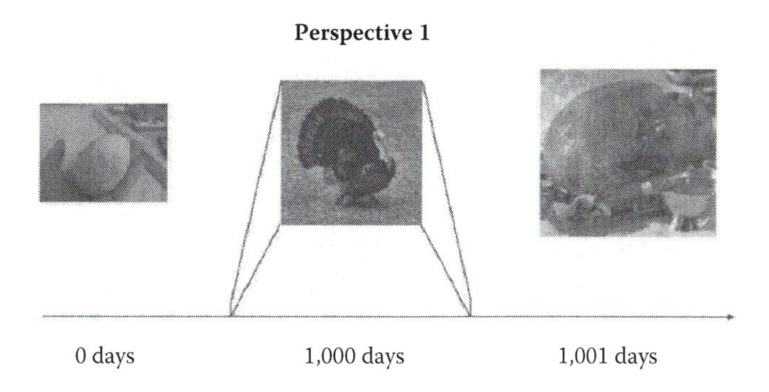

FIGURE 6.2 Expanding "turkey" perspective beyond 1,000 days.

world would be to understand why the turkey couldn't see outside of its box and take steps to change the perspective accordingly. Maybe the perceived impossibility of induction can be practically addressed by at least obtaining information beyond the localized experience.

Let's imagine for a minute how our turkey—if he had a human brain—could change his fate by changing the narrow-minded outlook fixed exclusively on getting food regularly. Had the turkey observed life on the farm for one thousand and ten days, he would have seen other turkeys being slaughtered, or at least disappear on their 1,001st day (Figure 6.2). This observation by itself would not save him from death, but the events would alert him to a potential anomaly lurking in the farmer's behavior.

Another way would be to rise one or two levels above the farm and see it as only one part of a larger process involving the whole food industry: the egg incubator, feeding farm, meat processing plant, bird feed manufacturing, food delivery businesses, supermarkets, and many other elements of the supply chain for feeding humans.*

Finally, to plan an escape, the turkey would have to know the farm inside out: the schedule, layout, holes in the fence, possible escape routes, and many other important details. By putting all these perspectives together—from a high-level overview to a detailed on-the-ground knowledge—the turkey might have been able to escape his predicament.

Of course, all these perspectives are beyond the brains of (contemporary) turkeys. Nevertheless, with some imagination, we can develop useful scenarios. Some of the "new turkey" vantage points described above were used in *Chicken Run*, a highly entertaining animated film showing how smart chickens managed to escape a certain death on the farm. With our much larger brains, we should also be able to change our perspective, identify the

* For an example of how to create a higher-level perspective, see http://xkcd.com/1110/.

right problem, and find the right solution. And in the process, we must break the trade-off between being too broad and too narrow with our problem definitions.*

Clear-sightedness never guarantees perfect vision because larger-scale developments affecting our future can be happening outside of our field of view. That is why we are likely to miss important opportunities if we go straight from the problem to the solution, bypassing a systematic effort to change the perspective. When teamwork is involved, such effort is also necessary because people often use the word *problem* to describe different states of affairs.

One type of problem we will call a *problem situation*. A problem situation is a general state of discomfort with the present and a feeling that something needs to be done to change it. For instance, when a patient comes to the doctor complaining of chest pain, it is a problem situation that needs to be resolved. At this point the doctor does not know what causes the pain, whether it is a benign or life-threatening condition. Depending on the urgency of the situation, she might perform an exam, order tests, and do everything necessary to transition from general symptoms to a specific diagnosis.

A second type of problem we will call a *problem task*. A problem task is a state of affairs where we know exactly what needs to be done to change the situation. In the heart patient example, that might be a surgery performed by highly trained staff to fix a clogged artery. Now the problem is clearly defined, and everybody in the operating room knows his or her role in dealing with the problem. With this second type, most of the problem-solving actions are executed through routine processes. In some situations, we can consider a specific problem task as a puzzle because the solution has already been tried and tested within a larger context.

To become successful inventors and innovators we need proactively and systematically to create various perspectives, understand the problem situation, and identify specific problem tasks that can resolve the problem situation. Our goal is to defeat the what you see is all there is (WYSIATI) bias and find scalable innovation opportunities.

---

* See the earlier discussion of the brainstorming tradeoff in Review Questions after Prologue, Soft Barrier 1 subsection.

7 Seeing the Outlines of the Box

Discovering the Boundaries of a System

In Section I we made the first step in the process of thinking outside the box when we started considering certain familiar devices as functional elements of a larger system. This simple step allowed us to broaden our vision of the situation and engage in divergent thinking.* For example, by looking beyond Edison's light bulb, we discovered his other important inventions that were critical to the success of the device itself and the electricity distribution system as a whole. More formally, whether working alone or in a group, we would like to answer the following questions:

- Assuming the entity under consideration is an element of a larger system, what functional role can it play within the system? Can it play more than one role at different times?
- Is it an instance of the Tool, Source, Distribution, Packaged Payload, or Control?
- Can we identify other elements of the system? How are they implemented? How do they interface with each other?
- What kind of element aboutness do we need to know to orchestrate the function of the system as a whole? What has to change in the system, so that it improves its performance as we add lots and lots of instances of elements, especially Tools, Sources, and Packaged Payloads.

At the conclusion of this question and answer session, we will have a system diagram of the present and a set of brief scenarios for the future. Even a simple system diagram of the present can uncover hidden assumptions and expose flaws in our current thinking. For instance, looking back at the

* Divergent thinking is a thought process or method used to generate creative ideas by exploring many possible solutions (http://en.wikipedia.org/wiki/Divergent_thinking).

Apple versus Nokia example in Chapter 2, we can see that Nokia engineers and managers focused all their attention on the phone—the Tool—ignoring systemwide changes brought about by the iPhone. In 2006, several months before Steve Jobs unveiled the iPhone, Nokia merged its fledgling smartphone operations with the highly profitable basic phone business unit, putting executives from the latter in charge. As the result, "The Nokia bias went backwards," said Jari Pasanen, a member of a group Nokia set up in 2004 to create multimedia services for smartphones. "It went toward traditional mobile phones," he added [9]. Even when Nokia did perform a detailed evaluation of the iPhone, including a complete teardown, they still saw it as an imperfect Tool within the box of the old communication system. Indeed, the iPhone was expensive and fragile when compared to standard phones at the time. But Nokia's narrow "clear-sightedness" prevented the company from seeing the iPhone as an element in a new system created by Steve Jobs and his multi-industry coalition.

At this stage of analysis we have to remember that in successful, scalable systems, the elements (e.g., Tools and Packaged Payloads) are present in a large number of instances. The key to Edison's breakthrough was his desire to produce and install not one, but lots and lots of light bulbs. The inventor had to develop an efficient way to deliver electricity to all of them. As the result of this insight, Edison created his remarkable parallel connection scheme. From that point on, finding a long-lasting filament for the light bulb became a specific problem task that he solved by a "99 percent perspiration" effort. Furthermore, Edison anchored his business model not only on selling the light bulbs, but also charging customers for electricity. The more power they used, the more money his company made. The need to know how much and when the customers needed electricity drove Edison's inventions of electrical power meters. Imagining lots and lots of instances of functional system elements interacting inside the system helps one imagine how scalable and successful the system can be.

In Section III we will consider in greater detail how the elements evolve within the system, how various problems show up in the process of system evolution, and how to converge on specific elements and interactions to solve the problems. In the meantime, to make sure we don't miss a valuable innovation opportunity, we will continue with the process of divergent thinking.

Our initial step was *horizontal*. That is, our mind's eye stayed on the same level, trying to figure out what other elements comprise the system and how easy it was to add instances that promote system growth. For example, we drew a diagram of Edison's system, moving from the light bulb (the Tool) to parallel connections (the Distribution), and other elements that make up the system.

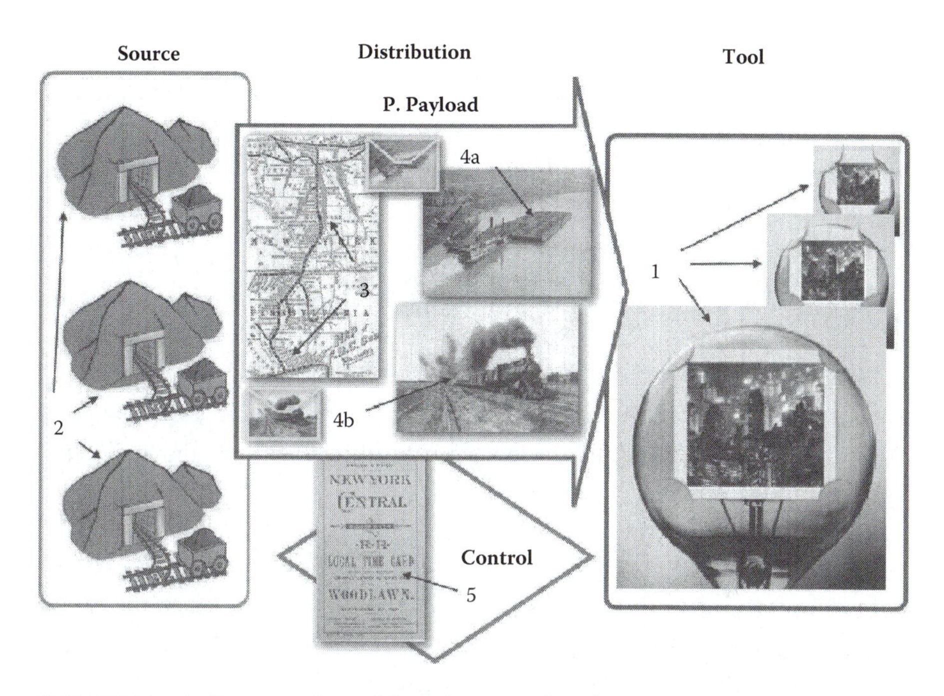

FIGURE 7.1 A diagram of a coal-based energy distribution system, with electrified Manhattan plugged in as an instance of the Tool. (1) Various city blocks "plugged" into the larger system (instances of the Tool). (2) Coal mines (instances of the Source). (3) Railroad routes for coal delivery (an instance of the Distribution). (4a) A coal barge (an instance of the Packaged Payload). (4b) A coal train (an instance of the Packaged Payload). (5) Railroad time table (representing an instance of the Control).

Our next step is *vertical*. We "level up" and consider the entire system outlined in step 1 as an instance of an element in an even larger system. We ask ourselves, what role does this element play within such larger system? After that we repeat the first horizontal step, but at the new, higher level.

For example, Edison's entire electricity distribution network can be thought of as an instance of the Tool, or one giant light bulb roughly the size of a couple of Manhattan city blocks. That is, the Tool converts coal into light. We need to find out what this giant light plugs into and whether it can play roles other than the Tool.

The diagram in Figure 7.1 shows our hypothetical higher-level system. Usually, we start identifying specific elements by focusing on the Packaged Payload first, because it represents essential ingredients produced, transferred, directed, and transformed by the rest of the system elements. The diagram shows coal trains and barges as instances of the Packaged Payload. The giant light bulb receives loads of coal and converts them into light. Facilities for unloading the coal and putting it into steam engines serve as an interface between the Distribution and the Tool.

The Distribution also delivers the Packaged Payload to factories, residential heating markets, and coal gasification plants, where coal is used to produce gas for gas lights. The instances of the Source are the coal mines connected to the railroads. The Control functionality for coal distribution is implemented through railroad companies that run delivery schedules, with the telegraph playing an important role in coordinating coal movement.

From this bigger perspective, we can recognize the opportunity for betting on electricity as the next generation of the American energy backbone. That is, instead of moving heavy trains over large distances, we can move electric currents. Transaction-wise, having electricity as the Packaged Payload would be much more efficient because it takes less energy to move electric current than an equivalent amount of energy in the form of coal. Once this conceptual-level substitution becomes technologically and economically feasible, it has the potential for dominating the next generation of industrial, residential, and commercial infrastructure. Engineer and industrialist George Westinghouse, whom we met in Section I, with his background in large-scale railroad business, appreciated this great opportunity and invested his time, money, and efforts to bring it into reality.

In Edison's world, the existing high-level system could grow by adding more railroad routes and more coal trains or barges. In sharp contrast, in Westinghouse's world the new high-energy electricity lines would bring energy to entire communities, replacing the coal trains altogether. In that new system, the transformer, or a collection of transformers, would play the role of a socket where the giant light bulb of Edison would fit right in. Other instances of the Tool can be plugged into similar sockets as well: factories, farms, supermarkets, communication stations, neighborhoods, and many more. To produce enough electricity on this new scale, large power stations, either coal or hydroelectric, would be needed to play the role of the Source.

Like Edison, Westinghouse saw the entire system. Unlike Edison, he saw it at a new, much larger scale. That is why he pursued a range of system-level improvements: from the transformer, to power lines, to generator turbines, to AC motors. Although as an individual inventor he was not as prolific as Edison, he did succeed with his vision by recruiting Nikola Tesla, as well as other engineers and inventors. The Battle of Currents that Edison and Westinghouse fought in the late nineteenth century decided not only whose electricity company was to prosper, but what the worlds' energy infrastructure was going to look like in future decades.

MOVING BEYOND THE BOX

One take-away message from the previous chapter is that the *vertical* step helps us discover large-scale innovation opportunities. The scalability problems created by Edison served as an opportunity for George Westinghouse, who envisioned an improved, highly scalable electricity distribution system. Westinghouse also knew that to take advantage of the opportunity, the vision alone would not be enough. That is, a move into the future at a higher system level requires large investments and takes many years to complete. Despite the significant advantages offered by electricity, it took 30 years (between 1900 and 1930) for just one city (Chicago) to become electrified—with the share of factories running on electric power growing from 4 to 78 percent [10]. Furthermore, according to UC Berkeley economist Brad DeLong, "Eight percent of American households were wired for electricity in 1907; 35 percent were wired by 1920; 80 percent were wired by the beginning of World War II" [11].

Understanding what the infrastructure should look like in one possible distant future helps us innovate and profit from the specific, lower-level technology inventions that must take place. Within the new, large-scale system, inventions that once were considered useless can become extremely valuable. For example, when Nikola Tesla joined Thomas Edison's lab in 1884, his ideas for the AC polyphase motor were not appreciated at all. In Edison's world there was no place for a large-scale electric power distribution infrastructure that could take advantage of Tesla's novel design. After a series of conflicts over money and technology vision, Tesla resigned from the lab and in 1886 founded his own company. However, not until 1888, when he started working with George Westinghouse, could his inventions find validation and the financial and technological support among the high-tech community of that time. The AC motor is but one example of many inventions that came before the real need for them emerged. In the next chapter we will consider how large-scale, long-term shifts in infrastructure form the foundation of "inventor's luck."

REVIEW QUESTIONS

1. Please choose an analogy that best mirrors the relationship between two innovators and the products and services they built.

   Edison is to Westinghouse as:

   1. Steve Jobs (Apple) is to Bill Gates (Microsoft)
   2. Steve Jobs (Apple) is to Larry Page (Google)
   3. Bill Gates (Microsoft) is to Larry Page (Google)

4. Mark Zuckerberg (Facebook) is to Bill Gates (Microsoft)
5. Brad Fitzpatrick (LiveJournal) is to Mark Zuckerberg (Facebook)

Sketch out a two-level system diagram and explain your choice.

2. In 1984, Leonard Bosack (age 22) and Sandy Lerner (age 19), both of Stanford University, founded Cisco Systems, now a multibillion-dollar global technology company that sells networking and communications products and services. Originally, the couple designed, developed, and sold routers for local area networks (LANs). Their primary buyers were universities and companies that wanted to interconnect various workstation computers on campus. Universities also used the routers to connect their networks to ARPANET, the world's first data network that connected most research and government institutions in the United States.

   What system element did the router represent: the Source, Tool, Distribution, Packaged Payload, or Control? What technological and business developments of the late 1980s and early 1990s helped Cisco Systems to scale their business up more than a thousandfold: from $200K to $2.2B/year in 1999? Sketch out a two-level system diagram and explain.

8 Inventor's Luck
A System Perspective

success = talent + luck

great success = a little more talent + a lot of luck

—Daniel Kahneman, psychologist, 2002 Nobel Laureate in Economics [12]

[F]or the first fifteen years after sliced bread was available no one bought it; no one knew about it; it was a complete and total failure.

—Seth Godin, entrepreneur and author of eleven books on marketing methods [13]

The common expression "the greatest thing since sliced bread" would make little sense to people before the early 1930s, the years when electric bread-slicing machines and wrapped, sliced loaves of bread started turning up in bakery shops in the United States. Suddenly, after more than fifteen years of development, the invention became "lucky" and enjoyed widespread commercial success. How did it happen? Why was there such a long delay in technology adoption? Was it due to, as marketing guru Seth Godin claims, the inventor's inability to publicize the idea, while paying too much attention to the technical implementation and the patents? To answer these questions, let us examine the actual story of Otto Frederick Rohwedder (1880–1960), the inventor of the original bread-slicing machine [14].

Why sliced bread? Because freshly baked or toasted bread tastes better to most people [15]. Unfortunately, a large, bulky loaf that is easy to bake cannot be later reheated thoroughly in contemporary ovens without burning the outside. If we heat the entire loaf at once, the part we don't eat quickly hardens and becomes inedible. To solve the problem of leftover bread, people would buy it in loaves, slice it at home as needed, and then toast the slices.

As a value-added service, bakeries could slice bread for their customers. Unfortunately, if the bakeries sliced the bread ahead of time, they would have the pain of having a large inventory of sliced bread go stale quickly (due to the higher exposed surface area). To avoid this loss, bakeries would slice bread manually, on demand, for some customers. This would take some time and effort, and still leave the customer with the problem of sliced bread going stale quickly.

At the turn of the twentieth century, electricity becme widely available. A number of inventors began thinking of ways of using it in all manner of domestic, industrial, and retail contexts. Otto Rohwedder started working on the idea of an electric bread-slicing machine in 1912. In 1916, he sold his three jewelry stores and devoted his full time to the project. In 1917, his enterprise suffered a major setback (bad luck!) when a factory fire destroyed blueprints and prototypes of the machine.

While making a living as a financial agent, Rohwedder continued working on his inventions. In 1927, fifteen years after starting the project, he came up with a machine that was able to not only slice, but also wrap bread into paper. This important modification solved the critical problem of sliced bread becoming stale more quickly than unsliced bread.

In 1928, Rohwedder sold his first practical device that included both the slicing and the wrapping capabilities to the Chillicothe Baking Company. In 1929, just as the sales of sliced bread began to pick up, the stock market crashed (bad luck!). To stay afloat financially, Rohwedder was forced to sell his patents to Micro-Westco Co. of Bettendorf, Iowa.

In the meantime, consumer demand for sliced bread continued to grow. In the beginning of the 1930s, motivated by the growing market for sliced bread, and taking advantage of the bad luck suffered by Rohwedder, the Continental Baking Company entered the market with its Wonder Bread, a presliced, prewrapped form of bread. Wonder Bread used a machine of its own design (more bad luck for Rohwedder because his patents didn't cover it), and by 1933, sales of sliced bread outpaced those of unsliced bread. A twenty-year quest from the original idea of a bread-slicing machine finally succeeded in becoming a practical innovation that enjoyed widespread adoption.*

According to Seth Godin, had Otto Rohwedder been more proactive in his marketing efforts, sliced bread would have become a commercial success much earlier, thereby enriching him, instead of the Continental Baking Company [13]. Or would it?

The answer is *no*. A reliable bread-slicing machine could not be produced before the late 1920s because reliable, inexpensive small electric power motors were not available at the time. And even if the machines were produced in volume, the demand for sliced bread prior to the second half of the 1920s was low. Consumers simply did not want much of it. At the time, toasting bread

* In system terms, a change in the Packaged Payload suddenly enabled the Source to dramatically scale up production and to take over the market.

was a rather tedious operation that produced low-quality results (mostly due to uneven heating, burning, etc.). Rohwedder and his would-be competitors in the sliced bread business were bound to be unlucky, because there were no standard toaster slots into which bread slices could be inserted in large quantities. That would all change after Charles P. Strite invented the Toastmaster, a slot toaster with a timer.

WHAT IS LUCK?

According to the online *Merriam-Webster Dictionary*, luck constitutes "events or circumstances that operate for or against an individual."

The circumstances seemed to operate against Otto Rohwedder. He suffered a devastating fire in 1917, lost his financial assets in the great market crash of 1929, and the competition overtook him when the market for his invention began to grow. Although he was very talented and steadfast in pursuing the idea, he didn't achieve great success on the scale of Wonder Bread.

On the other hand, there was a lot of luck for his *inventions*: sliced bread and the bread-slicing machine. What were the circumstances that worked for those inventions? Can the system model help us discover the origins of this luck?*

As we alluded to in the foregoing discussion, a brief system analysis (step 1) tells us that sliced bread is an instance of the Packaged Payload. That is, in the bread Distribution system it represents a new way to package a loaf of bread. Then, a bakery equipped with the bread-slicing machine becomes the Source. Like a power station that generates lots of electricity, the bakery generates lots of sliced bread. But where are the large numbers of Tools to use this new Packaged Payload? Why couldn't the Tools use the old hand-sliced bread as before?

We find the answer in the design of the toaster, where most consumers would "plug in" their bread slices. The modern toaster has a standard slot that works best with a bread slice of standard thickness. Although the standard slot seems obvious today, the original electric toasters didn't have it. Neither did they have a timer.

In 1909, GE introduced the first commercially successful electric toaster, which had two hot surfaces for sliced bread (Figure 8.1). Not surprisingly, given GE's heavy promotion of the Edison socket, users had to connect

* As an aside, a common tendency of people to focus on the individual inventor, rather than his/her inventions, tends to impair their ability to examine innovation from the system perspective. The inventor may—and often does—fail, while the inventions themselves can live on and succeed.

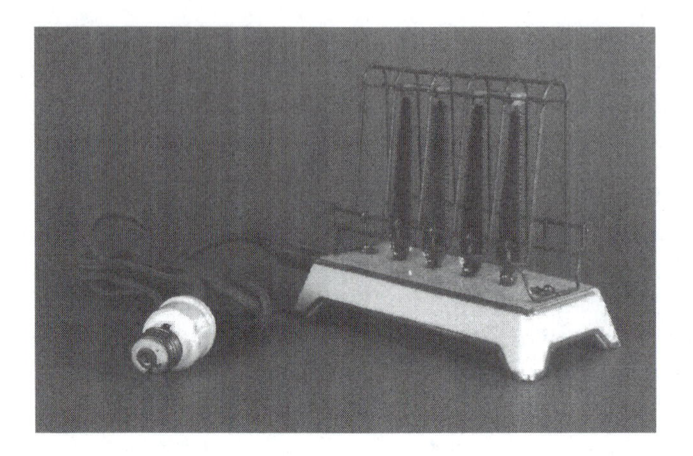

FIGURE 8.1 The original design of the early GE toaster. (Courtesy www.toaster.org.)

the toaster into an electrical power outlet using a cord that had a screw-in light bulb-like plug. (Thirty years from now, the USB wire connector for the iPhone will probably look as ridiculous as the light bulb plug on the toaster.)

With this piece of toaster history in mind, the beginning of Rohwedder's quest for sliced bread in 1912 doesn't look accidental anymore. Successful toasters were going to be "hungry" for sliced bread. Otto Frederick Rohwedder, although a jeweler by trade, was talented enough to recognize a big change coming into American homes. The future was going to be electric!

Let's investigate the toaster thread further. In 1920, Charles P. Strite invented the pop-up toaster with a timer (Figure 8.2). His design addressed two key problems inherent in the original toaster produced by the General Electric Company:

- Bread slices got burned if you forgot to turn them over in time.
- Bread toasting was slow because the heat was applied to only one side of the slice at a time.

By 1926, Strite's toaster turned into a major commercial success (Figure 8.3) when a consumer version became available (see instructions in Figure 8.4). By 1930, one million toasters were sold annually. It is no wonder that the sales of sliced bread started picking up around the same time. People loved using the electric toaster, but cutting soft, fresh bread into standard slices was a rather impossible task for an ordinary human. In 1943, when during World War II

FIGURE 8.2 Contemporary Toastmaster advertisement. (Courtesy www.toaster.org)

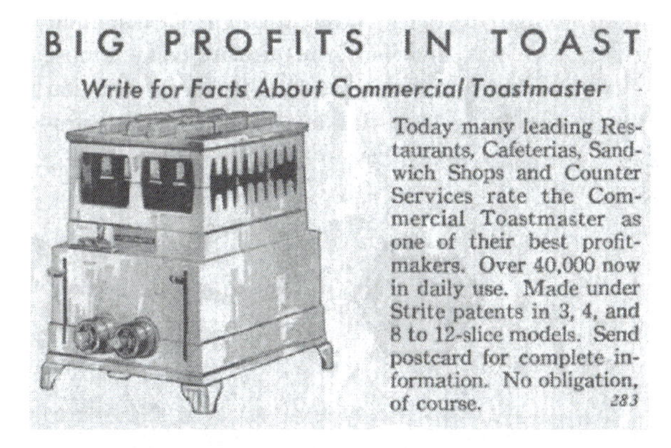

FIGURE 8.3 Contemporary advertisement for a commercial Toastmaster. (Courtesy www.toaster.org)

New York authorities banned sales of sliced bread due to perceived waste of packaging paper, one distraught homemaker complained that cutting by hand 22 slices of bread for her family's breakfast could threaten "the morale and saneness of a household."

To summarize, the bread-slicing machine and the toaster became elements of a growing system. First, the toaster created lucky circumstances

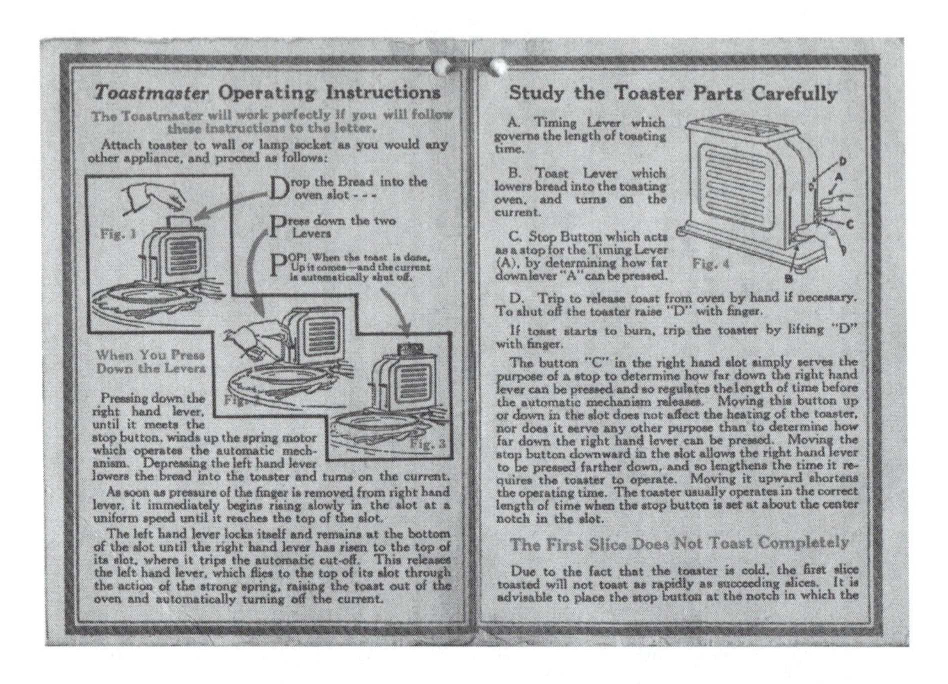

Toastmaster Operating Instructions

The Toastmaster will work perfectly if you will follow these instructions to the letter.

Attach toaster to wall or lamp socket as you would any other appliance, and proceed as follows:

Drop the Bread into the oven slot - - -

Press down the two Levers

POP! When the toast is done, Up it comes—and the current is automatically shut off.

Fig. 1

Fig. 3

When You Press Down the Levers

Pressing down the right hand lever, until it meets the stop button, winds up the spring motor which operates the automatic mechanism. Depressing the left hand lever lowers the bread into the toaster and turns on the current.

As soon as pressure of the finger is removed from right hand lever, it immediately begins rising slowly in the slot at a uniform speed until it reaches the top of the slot.

The left hand lever locks itself and remains at the bottom of the slot until the right hand lever has risen to the top of its slot, where it trips the automatic cut-off. This releases the left hand lever, which flies to the top of its slot through the action of the strong spring, raising the toast out of the oven and automatically turning off the current.

Study the Toaster Parts Carefully

A. Timing Lever which governs the length of toasting time.

B. Toast Lever which lowers bread into the toasting oven, and turns on the current.

C. Stop Button which acts as a stop for the Timing Lever (A), by determining how far down lever "A" can be pressed.

Fig. 4

D. Trip to release toast from oven by hand if necessary. To shut off the toaster raise "D" with finger.

If toast starts to burn, trip the toaster by lifting "D" with finger.

The button "C" in the right hand slot simply serves the purpose of a stop to determine how far down the right hand lever can be pressed and so regulates the length of time before the automatic mechanism releases. Moving this button up or down in the slot does not affect the heating of the toaster, nor does it serve any other purpose than to determine how far down the right hand lever can be pressed. Moving the stop button downward in the slot allows the right hand lever to be pressed farther down, and so lengthens the time it requires the toaster to operate. Moving it upward shortens the operating time. The toaster usually operates in the correct length of time when the stop button is set at about the center notch in the slot.

The First Slice Does Not Toast Completely

Due to the fact that the toaster is cold, the first slice toasted will not toast as rapidly as succeeding slices. It is advisable to place the stop button at the notch in which the

FIGURE 8.4 Toastmaster operating instructions. (Courtesy www.toaster.org)

for the slicing machine. Then, the machine helped the toaster become an essential household appliance. As part of the same sliced bread distribution system they turned into a consumer success story around 1930.

Note that our first, horizontal step of system analysis helped us uncover the relationship between sliced bread production (the Source), packaged bread (the Packaged Payload), and the toaster (the Tool). When we added an innovation timeline to the relationship, it became clear that marketing bread-slicing machines fifteen years before the toaster-related "window of luck" opened up would have been rather useless.

On the other hand, the introduction of Wonder Bread in 1933 made a lot more sense because it allowed the Continental Baking Company to ride a rapidly expanding system. Who knows, if Rohwedder hadn't been ruined financially in the crash of 1929, he might have become hugely successful in the wake of the sliced bread revolution. A large corporation can survive if it misses a window of opportunity by a year or two. For a small entrepreneur, timing his or her innovation is critically important. Even an easy, but systematic horizontal step outside the box can provide a much needed perspective advantage over the competition.

WHAT IS A LOT OF LUCK?

Although the first move in our system analysis process proved to be quite informative, let's not stop there, and let's make a second, vertical ("level up") move. We initiate it by asking the questions, "How could the system with lots of toasters and bread slicing machines become hugely successful as a whole? What makes it lucky?" Rather than focusing on the design of the machines themselves and Otto Rohwedder's bad luck, we choose to look at the large-scale system of electricity distribution that provided the environment for these appliances to succeed or fail.

As we discussed earlier in the chapter, the first forty years of the twentieth century in the United States were the years of rapid electrification. In 1890, Edison's DC electricity system was suffering from overwhelming consumer demand. Due to the lack of electric current generation capacity and significant grid limitations, the consumer would have to get on a waiting list to install a light bulb.

In sharp contrast, in the AC systems originally promoted by George Westinghouse electric power flowed abundantly. By the 1920s, electric utilities in the Chicago area were offering free electric irons to subscribers, along with power services financing spread over two years at no interest. By 1923, the standard power plug (an interface), as we know it today, came into existence. The toaster still remained somewhat of a luxury, but one didn't need to get on a waiting list to plug it in.

In the home, the light bulb turned into the number one application of electricity, with the iron being a close second. Both of them used long-lasting, high-resistance tungsten filaments invented in Europe in 1904 by Sandor Just and Franjo Hanaman. In 1906, GE patented a process for mass production of tungsten wire for light bulbs. Engineers used the widely available wire to make heating elements for toasters—the lucky toasters!

In the early 1900s, Henry Ford began using electric motors in his automobile factories. With increasing adoption of mass manufacturing methods, the electric motor technology became reliable and inexpensive for industrial use. It is not a coincidence that successful bread-slicing machines had built-in electric motors. The lucky machines!

To summarize, from the second-level perspective, the entire sliced bread distribution system looks like an instance of the Tool plugged into a large-scale AC-based electric power system. Its rapid growth depended on the broadening availability of electricity and electricity-related technologies. As we move back and forth in time at the higher system level, we can see the emergence of other lucky electric appliances that we take for granted today: refrigerators, radios, dishwashers, sewing machines, washing machines, stoves, fans, heaters, air conditioners, elevators, and others. The appliances are mass manufactured in factories running machines, assembly lines, and tools powered by electric motors. This transition from an energy infrastructure based on moving coal to one based on moving electricity created a lot of luck for inventors and innovators who applied their talents to solving emerging high-value problems.

9 The Three Magicians
Tools for Flexible Thinking

Our experience in applying and teaching system thinking methods shows that in the beginning they may feel too abstract. In a collaborative innovation environment, untrained participants might get confused and be unable to contribute at the top of their abilities. To address this problem, we can use the Three Magicians technique, which applies system-based concepts in a concrete, graphical mode. The technique originated in system-oriented methods for teaching storytelling to kindergarteners [16, p. 6], who have yet to develop abstract thinking skills. Furthermore, the Magicians helped kids overcome additional inherent biological limitations of humans.

Human short-term memory can hold and manipulate only three to four independent concepts simultaneously [17]—a critical constraint on our creative thinking. Unless we make a conscious, deliberate effort to analyze a

> Working memory capacity is severely limited in adults and infants, with both groups able to remember only about three separate items at once. One reason that adults are rarely conscious of this constraint is that we can hierarchically reorganize the to-be-remembered stimuli, thereby increasing the total number of items we can store. For example, the letter string PBSBBCCNN is much easier to recall after recognizing the three familiar television acronyms PBS, BBC, and CNN that comprise it.
>
> For several decades, this limited number was thought to be 7 ± 2. However, more recent analyses show that 7 ± 2 overestimates working memory capacity. When measures are taken to block chunking, adults can store only three to four items. This limit applies broadly to visual and auditory entities including colored shapes, oriented lines, spoken letters, and spoken words, for items presented either simultaneously or sequentially.
>
> **—L. Feigenson and Justin Halberda, "Conceptual Knowledge Increases Infants' Memory Capacity" [18]**

certain situation, we are likely to miss important details necessary for understanding its core problems. On the other hand, if we spend too much time dwelling on everything we know about the situation, we may never arrive at a creative solution because our imagination will be hamstrung by past experiences, a condition often described as *the curse of knowledge* or *analysis paralysis*. Ideally, we would like to have the best of both worlds: the big picture necessary for imagining a future *and* detailed knowledge for identifying and solving a specific problem task.

The Three Magicians approach is based on system thinking, which allows us to navigate between system levels at will, taking into account both the big picture and specific functionality details, if necessary. It also leverages our general psychological tendency to remember and recognize objects in space better than abstract concepts. This feature of the human mind has been exploited since ancient times for memorizing large quantities of information through a mnemonic device called the *method of loci* or the *memory palace* [19]. In this method, we first create an imaginary building with many different, but preferably familiar rooms. Then we walk through the building, placing the objects we need to recall later into the rooms and remembering how we got there. To recall the objects, we walk through the rooms and "collect" the objects [20]. Today, top performers use the method in memory competitions.

In the Three Magicians technique, the interrelated system boxes function as the building. Each magician helps us move in and out of the boxes when necessary. The magicians encourage us to look outside the building's windows and compare views from different levels.

- The first magician is Divide–Connect. His role is to provide a direct view at the problem situation. He helps us identify situation elements, see how they are connected, and which connections and interactions are involved in creating the problem.
- The second magician is Climb on the Roof–Lean to the Ground. He guides us in gaining different perspectives above and below the current focus.
- The third magician is Fall Back–Spring Ahead. His role is to take us back and forth in time, finding out how the situation, which previously was a recipe for success, could lead to additional success opportunities or turn into a recipe for disaster.

RUNNING A THREE MAGICIANS SESSION

The Three Magicians is a visual exercise that requires 30 to 60 minutes. Depending on the problem, the session can be done individually or with a multidisciplinary group of specialists. To free up participants' working memory, we ask them to draw all objects and connections on a whiteboard (Figure 9.1) or flipchart sheets. Research shows that for a visual exercise to be effective, images related to the problem situation have to be clear, interactive, and persistent.

In Figure 9.2 you see the layout of the visual workspace represented as 9 screens [21], each representing different perspectives in time and levels of zoom. We can also do the exercise using 9 sheets of paper, each representing a specific screen. Low-tech alternatives may include use of a flipchart with colored markers, a white- or blackboard with 9 or more areas arranged in 3 × 3 fashion. A high-tech option would be an interactive electronic video surface with zoom in/out and hand-drawing capabilities. No matter what technology we use to draw the images, it is important to keep all screens visible at all times. Erasing one screen to free up space for drawing another would violate the persistence principle and force participants to recall things, straining precious working memory. In the end, all the screens together, or at least the top 6 screens, should read like a movie storyboard, the movie telling a tale about a successful resolution of the problem situation. The

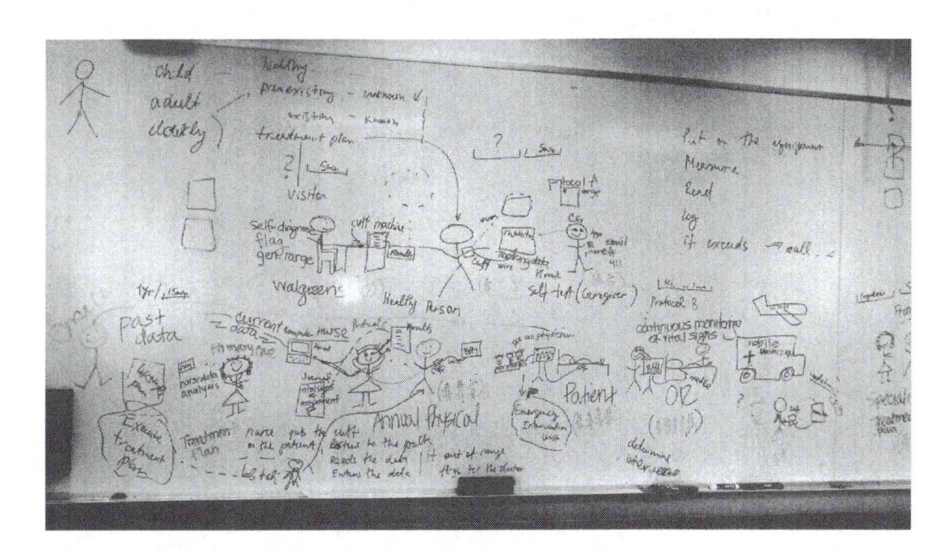

FIGURE 9.1 Whiteboard Divide–Connect sketches during a practice invention session for a novel blood pressure device. Stanford University Continuous Studies Program. Principles of Invention and Innovation (BUS 74), summer 2012. Photo courtesy Silvia Ramos.

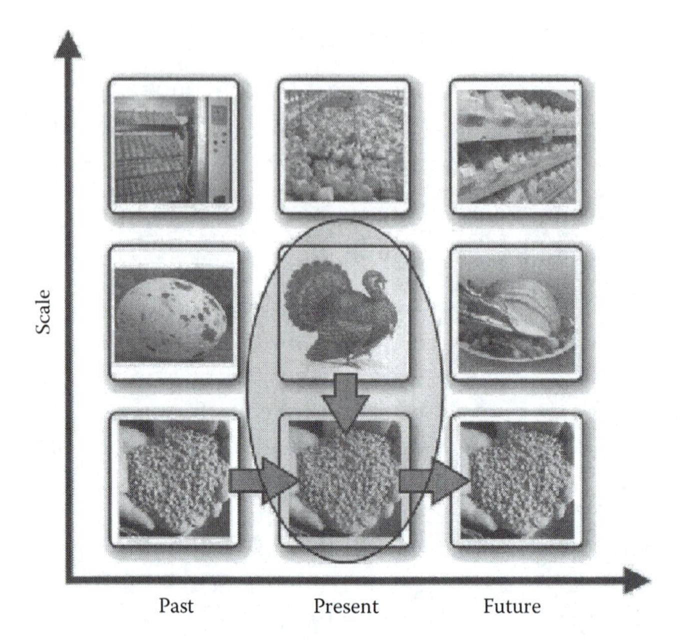

FIGURE 9.2 The 9-screen view. The turkey is preoccupied with the day-to-day supply of grain at the lower level. He doesn't see the bigger picture at the higher level.

storyboard should also show key problem tasks that have to be solved for a successful future to be realized.

The sequence of the magicians' work is very important. Figures 9.2 and 9.3 show how to avoid a typical inventor mistake when a problem solver attempts to think outside of the box by zooming in on a specific component, rather than zooming out to a higher-level perspective.

Similar to the system analysis we discussed in Chapter 8, the first step of the method is *horizontal*. Initially, we engage the first magician—Divide–Connect—and ask him to identify the box. To describe this approach in detail, we are going to reuse the War of Currents case study from Section I.

Magician 1: Divide–Connect (What Is the Functionality of the Box? Does It Work?)

Since Edison's most critical electricity-related project was to bring electric power to Manhattan, New York, we focus our attention on the town, with lights on the streets and in the buildings. The first magician draws everything he can see as he moves around. Because the area of the starting box is quite large, the magician has to observe city life from a bird's-eye view and make sure he doesn't miss any important detail. His main goal is to identify and draw key user scenarios; the more, the better.

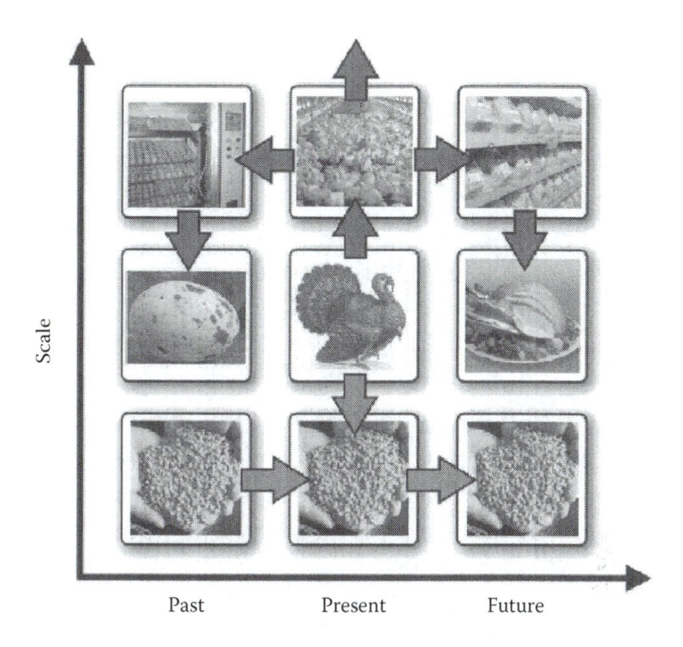

FIGURE 9.3 The 9-screen navigation logic. To understand the situation, the turkey needs to follow the top level arrows and see how the problem develops in space and time.

Year 1886, Manhattan. First, we notice electric lamps on lamp posts lining the street. It is getting dark and a man in a little switching station turns on the switch and the lamps light up the block. There are some houses, restaurants, stores, and offices with lights shining through the windows as well. Watched by a policeman, pedestrians are traversing the sidewalk. A pickpocket is hiding behind the corner of a dark residential building. We can see dim lights flickering—some people still use gas, oil, and kerosene lamps. A horse carriage drives by; the street is brightly lit, so it is not a problem for the driver and the horse to see the way. A fire brigade sits idly by the fire station. Since the introduction of Edison's electric lights, there are fewer fires in town and the idle firefighters gawk at the new street life. It is raining, but transparent covers protect the electric lamps from the elements. A cat on the roof of a restaurant watches a bird sitting on a lamp post. We notice that electric lamps are protected from bird droppings as well.

How does electricity get into the light bulbs? When we examine the base of the lamppost we see wires going into an underground conduit. As we follow it for no more than a mile, the conduit brings us to the switching and metering station (an often overlooked invention of Edison that is critical for the entire electric utility business model) where they keep track of how much electricity customers actually use. Some meters are located in the station, others on customer premises. A person records metering information and once a month sends bills to building managers, restaurants, and residents for the amount of electricity they used. Another person keeps track of requests for new light bulb installations and sends out electricians to install the bulbs.

The conduit leads further to a power station with a dynamo machine that generates the electric power for the lights. The dynamo machine is attached to a steam engine that noisily rotates its wheel, spewing black coal smoke into the air. A team of men shovel coal into the steam engine. Earlier that day, another team unloaded the coal from a train that came in via the railroad, delivering coal at regular intervals in order to drive the power station.

This quick walk through the scene shows us how things work inside the box. That is, the *box* is largely a self-sufficient system observed from a predefined perspective. In the example, it is a snapshot of life on a street equipped with electric lights. As we examine the insides of the box, we can see the benefits various participants derive from the lights: people work longer hours; they walk, shop, and visit restaurants after dark safely. Because of the extended light hours, the downtown businesses are booming; more patrons and revenue.

We can also see some negative effects, such as smoke and noise within a relatively small distance between a prime area in Manhattan and the power station. We also note that the steam engine is not very powerful, although it is big enough to generate power for a few city blocks. The coal may also present problems. To have a reliable supply of energy, the station has to be close to a railroad or riverbank, which are typical delivery routes for coal trains and barges. Finally, as the demand for light ebbs and flows depending on the time of day, steam engines have to be started, stopped, or operated unproductively, disconnected from the generator.

After we draw the first (central) picture, we check it for completeness. At this stage, we ask the magician just two questions, "What is the system in your focus supposed to accomplish?" and "Does it actually work according to your description?"

First, we can say that Edison's city lighting system is supposed to allow businesses to extend their work hours after darkness falls. It also reduces the fire hazard, adds to street safety, and increases security in a variety of weather conditions.

Second, the storyboard scenarios we drew on the whiteboard seem to work. Not necessarily for their artistic qualities, but because they help us trace production, distribution, use, and control of electricity throughout the story. Also, we can clearly identify the business model. The system makes money by charging for power use and selling electric equipment, including the light bulbs.

We can also see some opportunities for improvements, although it is far from clear that anybody needs them yet. Let's try to create a bigger picture before we attempt out-inventing Edison on his own turf.

Magician 2: Climb-on-the-Roof (Does It Scale?)

Now, we ask the second magician to rise above the first picture and report on views that lie beyond the one at the center of our attention. In this particular situation, we imagine going "on the roof" of New York City or riding a cloud that flies above it. (Neither airplanes nor large air balloons had been invented yet; otherwise we would've taken one of those). In the beginning, the second magician spots objects and interactions that the first magician could not see due to his limited perspective.*

As we walk on the clouds above New York and the East Coast, we notice that the majority of city blocks are still lighted by gas lamps and some don't have lights at all. Maybe those areas are too poor to afford electricity which, after all, requires a massive investment into power stations and underground electric distribution systems. Probably, local businesses can't generate extra income from the extended light hours yet. Or maybe they are too far from the river or railway and it is not practical to carry large amounts of coal there.

We also observe just one or two factories that use electric motors. Some motors are driven by steam engines, others by waterwheels. However, the vast majority of factories do not appear to be using electricity at all. They have an ample supply of coal, but it is used for powering steam engines that drive various pieces of factory equipment. When we peek into a factory window, we see that each engine is set up to distribute its mechanical power to other machines through an intricate system of levers and pulleys. Wouldn't it be nice to replace all this mechanical complexity with just wires and electric motors?

What if instead of bringing coal to the city, we bring electricity itself? What if we built one giant power station that uses either coal or water energy to produce vastly greater amounts of electricity than the small power stations designed by Edison? Could we then bring electricity to the factory and distribute it in any way we want? We could do the same for city blocks that do not have easy access to a river or a railroad. Rather than expanding the railroad into those regions, we could bring a lot of electricity from afar and

* For a graphical example of using a "second magician"–like perspective, see http://xkcd.com/1110/. In a nonfiction example, Charles C. Mann applies a similar change of perspective in his book *1493: Uncovering he New World Columbus Created*. To show the impact of silver metal deposits discovered in South America on global trade, the author rises above cities, continents, and oceans and describes how coins stamped out in Potosi (now Bolivia) flow to Spain and China in exchange for goods, people, and services [22].

distribute it in smaller chunks to those who could use it, either for running factories or lighting streets, houses, and offices. Some people think the new power distribution system could even replace horses and drive an electric city tram (see, for example, US Patent 295454, Sprague Electric Motor, 1884).

This might be a great idea, but electrical engineers are telling us that with Edison's direct current (DC), such long-distance delivery would be practically impossible. DC at the time is not suitable for adjusting electric current from one source to a wide variety of specific needs of motors and light bulbs. To implement our idea in a commercial environment, we would have a hard time taking a large amount of electricity from the super power station and reducing it down to different usage requirements. It appears that if we wanted to use electricity on a larger scale and for a diverse group of users, Edison's solution would not work at all.

Note what just happened. With the help from the second magician we found important scalability problems that Edison's DC system could not address. The system has two fundamental limitations. First, it generates and uses electricity locally only. This means that coal has to be brought deep into the city by train or barge to feed steam engines. Second, to generate and distribute electricity at a location that requires a lot of power, we would have to build dedicated power plants and dedicated electric lines, because matching currents to a wide variety of user needs is impractical in a centralized DC-based system. Despite all the talk about Edison's inventive genius, we can see that the door of opportunity to out-invent and out-innovate him remains open.

At this point, the key question that the second magician has to answer is, "Does the system under consideration scale up?" In Edison's case, the answer is *no*. Although a strong indicator for potential inventions, the answer does not necessarily mean that we will be successful in building a large-scale replacement for Edison's implementation just yet. Besides, his business model appears to be quite successful. His company, Edison Electric, builds as many local power stations and power lines as possible. His light bulbs are in high demand. For us to be successful in creating something ten times better, we would have to see whether there is a need for innovation at a larger scale. This is where the third magician comes into the picture.

Magician 3: Fall Back-Spring Ahead (Is There a Need for Scale?)

Staying at the level of the second magician, that is, walking on clouds, we ask the third magician to take us back in time; for example, one generation of technology earlier. We see the same map, but there are no city blocks with electric lights whatsoever. Some of the areas have gas lighting; others are totally dark at night. Ideally, the third magician provides us with a short time-lapse picture sequence that shows how things change over time, frame by frame, between now and the earlier state of the infrastructure. By running the movie back and forth, we notice that cities and industries are growing rapidly.

Ports, railroads, factories, banks, telegraph lines, and department stores are expanding. It looks like many places would benefit from electric power, even if the growth is not quick. More local businesses would prosper from extended hours. Additionally, electric lights are a lot safer than gas lights. Therefore, light conditions in offices and houses would also improve without an increase in the risk of citywide fires. For industrial applications, the invention of practical electric motors shows great promise in manufacturing and transportation.

As city populations continue to grow, horse-driven trams can't handle the demand and the problem of manure on the streets becomes more acute [23].* The steam engines everybody uses right now produce great power, but they are very heavy, dangerous, and require a lot of coal to be hauled in. The use of steam engines in factories involves an inefficient, complex mechanical system of line shafts and belts. Replacing it with electric wires and motors would save owners a lot of money through reductions in energy, maintenance, and labor costs.

While staying at the high level, let's spring ahead and try to extrapolate life outside of Edison's system. On the one hand, we can build more power stations in cities and factories to add light bulbs, electric motors, and electric trains. On the other hand, if we do that, we'd have to bring additional railroad lines into the city, allocate additional real estate for power stations, build them, and bring more coal. However, we would still be limited in terms of the location and the amount of usable electricity because the number of hungry, noisy, and dangerous steam engines would increase dramatically. A more practical alternative would be to bring electricity from places where coal and water energy are cheap. Building electric networks must be more efficient than expanding railroads and energy production into the city.

When the third magician shows us the Fall Back–Spring Ahead movie, we realize that there is a *need to scale*, which Edison's DC-based system cannot satisfy. In other words, a great innovation opportunity is available. A set of technological and business innovations would be justified in bringing electric energy over long distances at a low cost.

* In 1898, New York hosted the first international urban planning conference. The agenda was dominated by discussions of the problem of horse manure, because cities around the world were experiencing the same crisis. But no solution could be found. "Stumped by the crisis," writes Eric Morris, "the urban planning conference declared is work fruitless and broke up in three days instead of ten."

The First Council of Magicians: Lean to the Ground (Past and Present)

By now, you may have noticed that the magicians work together all the time. The first magician, Divide–Connect, is always present, helping us identify elements and their relationships. Initially, he does it at the city street level, but then working with the second and third magicians, he runs the same exercise from a higher-level perspective. That is, the second magician moves our perspective up and down in space, by zooming out and zooming in above and below the current focus of attention. The third magician moves the perspective back and forth in time, starting at the highest level.

During the exploration of the electricity system, the second magician made sure his partners stayed on the roof while moving back and forth and making the time-lapse sequence. But to understand the benefits of electricity compared to the old gas lighting system, the third magician fixed the time frame in the past, then the second magician instructed the first one to climb down to earth and explore the life of the same city street in the past. This is how we noticed the city fires caused by gas and kerosene lamps. We also saw piles of manure on the streets, dark buildings and businesses after dusk, inefficient steam engines at factories, and all the other problems. Because each magician is responsible for his own moves in the system's "palace," we can seamlessly navigate between the various boxes. Although the level of detail increases as we explore the problem situation in space and time, our working memory stays active because we keep our findings on the storyboard, not in memory.

The Three Magicians help us overcome a biological constraint imposed on human divergent thinking. They guide us through a systematic exploration of the problem situation, combining the benefits of high-level vision with close attention to detail, which is necessary to spot specific problem tasks. When used in multidisciplinary team meetings or brainstorming sessions, the Three Magicians technique helps all participants to contribute their professional knowledge and problem-solving skills.

REVIEW QUESTIONS

A Three Magicians exercise (30–45 minutes): Research shows that in the workplace, "same-day joy is associated with higher creativity, and same-day anger, sadness, and fear are associated with lower creativity." (Creativity is defined as "the production of novel and useful ideas." Same-day positive and negative emotions refer to the feelings one experiences on the day when he or she has done something creative. Creativity-related emotions do not carry forward to the next day) [24]. The goal of the exercise is to help you become happier and, if necessary, find ways to reinvent yourself or your organization.

Using the Three Magicians method, sketch out your everyday work or study experiences. Start with people, objects, and challenges you encounter, and note the nature of the interactions (Divide–Connect, the first magician). Where does your current situation belong on the Skill–Challenge diagram (see the Prologue, Figure 0.6.)?

Move to the upper level and sketch out regular processes and work environments (crowds) with which you are involved during the day (Climb on the Roof, the second magician). Are these processes and organizations creative or routine? (Please refer to the Prologue, "The Creative Crowd" section for the discussion of *creative* versus *routine*.)

While staying at the current level, move back in time several years and sketch out your previous work environments (Fall Back, the third magician). How does your current work environment differ from the previous one? Is there a trend over the past several iterations of work environments, or within one or two?

Go down a level (Lean to the Ground, the second magician) and place your old self on the Skill–Challenge diagram. How does your current position on the diagram differ from the previous one? Compare your old skills and challenges to the new ones. Which skills are getting better? Do new challenges emerge? If you feel you are becoming less creative, you may move down one more level and evaluate the specific skills–challenge combinations (personal, technical, social, professional, etc.). Which Skill–Challenge gaps are getting bigger?

Go up to the highest level (Climb on the Roof, the second magician) and move several years into the future (Spring Ahead, the third magician) until you start detecting a possible change in your work environment (processes). Is the environment becoming more creative or more routine? Will the environment increase or decrease your luck with regard to being creative and/or productive? Can you influence or change the environment to affect the luck factor? Sketch out your luckiest and unluckiest environments. What external trends (tailwinds and headwinds) could change the environment? What challenges would they create? Do you see any constraints that are going to break down? What alternative scenarios can you see available?

Go down a level (Lean to the Ground, the second magician) and evaluate your Skill–Challenge combination in the new environment. Are you a part of the creative crowd? How would you take advantage of the lucky situation or mitigate the unlucky one?

Go down one more level and identify specific skills and challenges that have to produce a better match. Describe your new, reinvented self. Develop an "innovation" plan to implement and experiment with your new invention. Set up milestones and commitment devices that could help you monitor your own progress and changes in the environment.

10 Imagination Development
Seeing the World beyond Present-Day Constraints

When we use system diagrams or the Three Magicians storyboards, we improve our skill to consider a complex problem-situation from new perspectives. This makes it easier to shake off the constraints on imagination imposed by various contexts we typically consider a natural state of affairs. Thinking beyond ordinary is essential for discovering a potential direction for a breakthrough innovation. Often, it involves playing with the parameters of a given problem situation, stretching the reality to the extreme. In this chapter we will introduce one more imagination development tool that helps both divergent and convergent thinking.

In March 2012, Amazon.com Inc. bought Kiva Systems, a company that develops robotic systems for warehouse automation, for $775 million in cash [25]. According to industry experts, this technology acquisition will enable Amazon to create a scalable platform for fast and cost-effective order fulfillment.

In a 2011 TEDxBoston talk, Mick Mountz, the inventor of the technology, explained his "beyond ordinary" thinking that led to the development of the breakthrough idea (Figure 10.1). He showed that traditional warehousing systems were designed for shipping pallets and cases of similar goods to large retail stores, such as Staples, Walmart, Costco, and others [26]. By contrast, the new reality of Internet commerce created a need for packing and shipping small, diverse orders to individual online shoppers. (In system terms that would represent a change in the Packaged Payload.)

When Webvan, Mick's previous e-commerce company, tried to use traditional fulfillment methods, handling costs escalated. For example, it cost the company one dollar to pack an 89-cent can of soup—a whopping 112% handling premium. Burdened by the costs, Webvan eventually went bankrupt. The failure made Mick think about a fundamentally different approach to e-commerce order fulfillment.

The breakthrough came when he started considering "what-if" scenarios that involved a totally different scale of operations, stretching to the limit the constraints of the existing model. He imagined a warehouse of infinite size

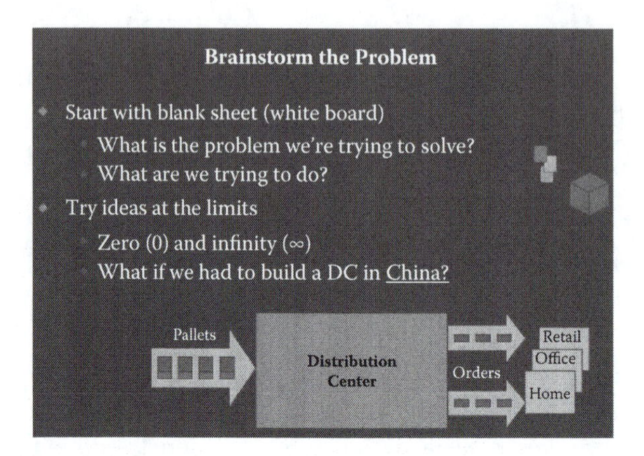

FIGURE 10.1 From zero to infinity. Mick Mountz's TED talk screen shot. (From Mick Mountz: Let the inventory walk and talk, *Ted Talk*, 2011, http://www.youtube.com/watch?v=szU2-1infqc)

with an infinite number of workers and zero dollars per hour in direct labor costs. In the old system, to fulfill an order the worker would run to a shelf, find the right product, and bring it to the packing station. One such worker described himself as a "warehouse slave," struggling to keep up with the goals given to him by the supervisor [27].

But in Mick's infinite warehouse, it would take forever to find the right product. To solve the problem, he imagined two separate types of workers: *runners* and *packers*. Each runner would hold his product and, upon request, would bring the desired product to the packer. As a result, the search operation would be eliminated completely. At zero dollars per hour, it would cost nothing to employ an infinite number of runners holding and delivering an infinite number of items. Essentially, the products would walk and deliver themselves to the packers.

To implement the idea, Mick and his colleagues developed a robotic warehouse distribution system, in which robots played the role of runners, delivering goods to packing stations (Figure 10.2). Guided by computers and communicating with each other via a wireless network, the robots would bring the right products to the right packer at the right time. In addition to eliminating search errors and in contrast with the traditional assembly lines, the new packing system allowed for complete independence between packers. Therefore, the number, location, and specialization of packing stations could be scaled up and down depending on the number of orders—a valuable feature for online shopping, where demand varies widely from season to season. A what-if, reality-stretching thought experiment led to an innovation that

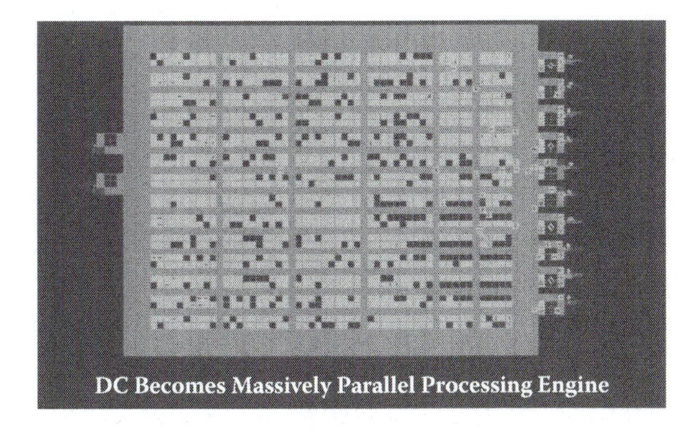

FIGURE 10.2 A bird's eye view of a large-scale, distributed robotic warehouse. Each dot repesents a mobile runner-robot that can hold and deliver a good to packing stations on the right. To reduce handling time, robots with popular items can stay closer to the packing stations.

changed the reality of a multibillion-dollar industry further validated by its acquisition by Amazon.

Although Mick Mountz's idea looks like a serendipitous discovery or an iteration on a previously failed business idea, we can use this approach to think deliberately and systematically *beyond ordinary*. When we work with inventors, we find that their professional intuitions are often confined to the scale of things and events they encounter within familiar professional contexts. This became apparent during the workshops that Eugene, one of the coauthors of this book, conducted with the SmartSurfaces classes at the University of Michigan* led by John Marshall (Art and Design), Karl Daubmann (Architecture), and Max Shtein (Materials Science and Engineering). There, working with very skilled students and seasoned practitioners from three different departments, he demonstrated that there is a natural tendency to focus thinking at the scale of typical professional experience. For example, some scientists specialize in designing and synthesizing nanostructures, objects measured in nanometers (10^{-9}m, which is 50,000 times smaller than an average human hair). When discussing potential applications of their amazing technologies, they typically have difficulty zooming out and translating novel properties of nanomaterials into the benefits for everyday consumer devices, which are several orders of magnitude larger in size than "ordinary" nanostructures.

In the other extreme are architects, for whom the design of a city or a building can involve tens of square feet or even miles of various structures. In turn, architects have trouble appreciating problems associated

* See http://www.SmartSurfaces.net for more detail on the project.

with integrating a practically infinite number of nanoelements into mile-scale living spaces. Industrial engineers and designers, who typically work with objects a human hand can touch and manipulate, operate somewhere between the two extremes. Neither the infinity of nano, nor the infinity of mile-size installations lend themselves to an easy assessment and utilization within the capabilities of their familiar mental tools. Amazingly, when they work together as an interdisciplinary team for a sufficient period of time, they become much more comfortable thinking outside of their "natural" professional experiences.

The purpose of this chapter is to reap the benefits of thinking beyond ordinary scale, where by using a simple imagination development technique that takes up 5–10 minutes prior to an invention session, we will be able to invent more and better. Just like athletes stretch before the game, inventors may want to warm up their imagination before taking on a difficult problem.

SCALE-TIME-MONEY (STM): TOOLS FOR EXPONENTIAL THINKING

The STM operator was originally created by Genrikh Altshuller to help problem solvers develop imagination skills through a series of playful exercises [28]. The first goal was to focus on functionality, rather than implementation. The second goal was to explore the system by exposing its limitations and discovering new features by radically scaling it up and down. This scaling can be done through common dimensions; size, time, and money (money being an easy proxy for resources necessary for implementing the functionality). One way to think about the exercise is to imagine yourself being Alice in Wonderland, where she tries different potions to shrink and expand herself. You may also imagine yourself a Mick Mountz, who stretched his warehouse to infinitely large dimensions, while the money needed to manage it becomes infinitely small.

Here is an example of how the STM operator might work. First, let's pick an everyday object on a scale we consider normal and think of its main useful function. For example, the laptop computer used to write this book runs software applications: a word processor, email, web browser, video player, game machine, and others. Now, let's take the size dimension and in our mind's eye start imagining an object that will have a similar function, but on a larger scale.

Consider what a computer the size of a whole house could do for us. Perhaps, it could entertain all family members, provide security, and control climate. Maybe it will also run applications that help us operate household robots, trim backyard plants, and zap pests. Clearly, the computer should comprise many *thinking* parts that talk to each other. Otherwise, it would take over the entire space of the house and push out all its human inhabitants.

Note that as we changed the scale, a new quality appeared. The house-computer no longer comprises a single device running different applications. Rather, it has become a dynamic set of different devices needed to run a vast array of applications around the house (where the term *software application* or *app* stands for, e.g., a household chore). Most of them need to communicate with others, depending on the situation. For example, adding sensors to the computer-home will increase the amount of information passed between the sensors and the decision-making devices. Also, the change in scale alters the frequency and complexity of interaction between people, pets, robots, devices, and the environment. The speed and reliability of the network inside the house becomes a critical component of any such application. Similar to Mountz's warehouse, communications and orchestration of activities would be essential to the performance of the new system as a whole.*

We can also imagine a house full of computers, that is, a data center. The applications it runs process and serve information to lots and lots of other computers. Each normal computer within the data center no longer runs independently. Instead, all of them have to coordinate user requests for shared data center resources. Accordingly, the problems that such a scaled device encounters differ substantially from the previously normal set. What if a rogue piece of software appears in the house? If on a regular laptop it can cause damage to one or a few persons at a time, in a data center it can wreak havoc on an exponentially larger scale, stealing money and corporate secrets, corrupting data, and inflicting losses to vast digital communities. Counteracting such a threat is extremely difficult. For example, when Google servers were infiltrated by hackers from one of the company's offices in China, the search giant chose to close a large portion of its operations in the country rather than risk similar attacks in the future.

Imagine walking into this computer. What do you see? Is it just a bunch of normal computers stacked together in neat piles? Unlikely. The new computer design does not need a keyboard, a display, or a mouse for every processor that runs software applications. Since this is just an exercise and you are not going to suffer the consequences, feel free to explore various parts; tug on wires, plug and unplug power, add and remove applications, break hard drives, and see how the "house" reacts. Observe the effects and develop ideas about new functionalities that the scale difference creates. For example, to save money in its startup years, Google engineers created a breakthrough when they used cheap, relatively unreliable hard drives considered unfit for

* Note, for example, that the warehouse becomes susceptible to novel software viruses we discussed in Chapter 3.

data centers. To make the system reliable as a whole, Google software would store data in multiple locations and retrieve it intact even when some of the drives failed.

Let's go further. Can you imagine a computer on the scale of a country or a whole planet? It should work well beyond the pages of the World Wide Web. If you are a sci-fi fan, some popular stories immediately come to mind. Solaris, the planet-brain with long-distance telepathic abilities in a Stanislaw Lem novel. *The Matrix*, a world-size computer harvesting energy from human bodies in the Wachowsky brothers' movies. Pandora, the sentient planet from James Cameron's *Avatar*. The Borg—a part-human, part-machine species that shares a collective intelligence and aimed to assimilate other species—in the series *Star Trek*, or perhaps the character Q, who inhabits a vast continuum and could travel across galaxies at will. The authors did not disclose how their creations work, but we could get an impression of their infinite sensing and computing power, which were well beyond human comprehension.

Would the planet-brain run on the same kind of processors as your laptop does? If so, how much energy will it need to generate and dissipate? If the processors were different, some with high, others with low power, how would it connect its own mind cells and coordinate information flows between them? How would it decide which streams of consciousness get priority in using its computing resources? What kind of signals and interfaces will it use to interact with the creatures that inhabit the planet? Will its memory be too slow to coordinate interactions between natural and artificial life? Maybe it will work by activating different layers of intelligence. Or maybe the planet-computer already exists and is making its baby steps when listening to humans talking to each other via Twitter.

In any case, a computing platform on this scale will be vastly different from any laptop or even the state-of-the-art existing data centers. Energy use, communication protocols, application principles, and many other aspects of normal computing will have to be rethought because they simply stop working on such a grandiose scale. For example, copper-wired connections may fail due to electrical resistance losses. Furthermore, given the enormous distance between elements of the system, signals traveling at the speed of light may not succeed in delivering messages. For example, in a truly infinite warehouse, Mike Mountz's robotic runners would have to become intelligent enough to anticipate orders rather than react to real-time delivery requests. (This last strategy—anticipation—is already being implemented by search engines and marketing firms, provided they have sufficient data about the originator of any given query. Of course, anticipation works as long as the anticipated events keep occurring.)

Let's change the direction and start shrinking the laptop. If we make it the size of a small coin, what functional abilities would we like to keep? Assuming the same density of computing elements inside the device, it would probably have a lot less general intelligence, but might be quite capable of doing certain highly specialized work instead. For example, it could run air conditioning in your car and monitor overall energy consumption. It could also talk to other cars and traffic signals to share the road and deliver desired performance, depending on your destination and route preferences. As we discussed earlier, it could also borrow intelligence from a larger computing system that has a better temporal and spatial perspective on higher-level processes. By combining various processing and communication functions within the same computing module, we can create reconfigurable modules that can adapt to changing internal and external conditions (see Figure 10.3, US Patent Application 20120007038. Reconfigurable Multilayer Circuit. Dmitri B. Strukov, R. Stanley Williams, and Y. Eugene Shteyn).

Since the tiny computer is now responsible for somewhat simple but vitally important parts of human life, we can't protect it from hackers with

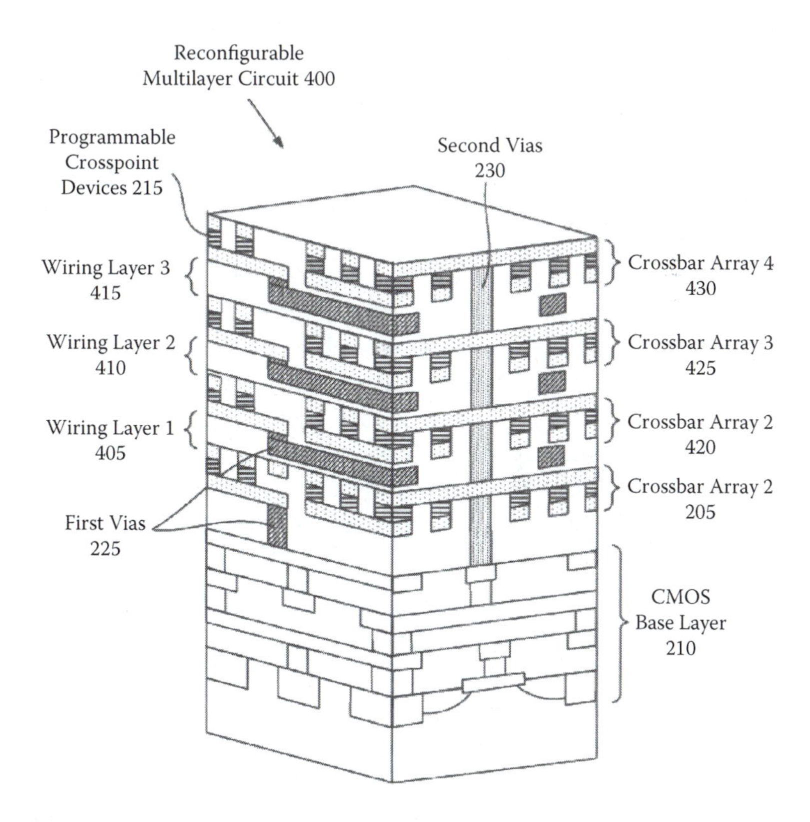

FIGURE 10.3 Reconfigurable multilayer circuit, US Patent Application 20120007038.

once-a-day scanning for predefined bit patterns adopted for laptops. When driving or operating industrial machinery, even a minute deviation from the right process can cause major damage. For example, Stuxnet, a mutating virus planted into computer chips that controlled Iranian nuclear fuel enrichment centrifuges, clandestinely destroyed many of them by causing erratic behavior of their electric motors. Even simple devices like a computer mouse, which has its own tiny processor, were found to host spyware programs that transferred all user keystrokes to a remote location. Similarly, a simple malicious program hosted by a driverless car can selectively target a rental agency, a car manufacturer, or an insurance company. These are just some examples of problems gaining in importance as computers shrink and become more interconnected.

Thinking on an even smaller scale, what could a computer the size of a sand grain do for us? What applications would it run? Should it work, for example, on biological principles, replacing the mass-market computing paradigm that dominates today's world? Let's say we would like to treat Type 2 diabetes, a lifestyle-related disease caused in part by unhealthy diet and low levels of physical activity. Would our tiny computers do the job of real-time monitoring of blood sugar levels and neutralize unwanted fat molecules? Or would it send signals to the brain requesting a higher level of physical activity, or perhaps wiser dietary choices from its human host?

What would it take to develop an auxiliary immune system that, instead of viruses, could detect and eliminate cancer cells, prevent Alzheimer's-related deterioration of brain cells, and monitor desired elasticity of our blood vessels? Clearly, today's principles of working with information implemented in regular chips will have to be modified to accommodate such tiny, smart particles. On the other hand, we already have nerve networks in almost every part of our body. Maybe we can *teach* them how to run new applications. Futurists often imagine and discuss such artificial intelligence (AI) applications, but with the STM operator, everybody can explore this innovation space systematically and come up with new ideas. The operator enables ordinary people to push their thinking beyond everyday constraints and discover new opportunities.

As we scale the system up and down, we begin to find that some connections and interfaces between elements fall apart, while others strengthen. For example, the smaller the element, the simpler its behavior, the more of them can be produced and deployed within even the tiniest space. But once you have so many tiny elements, packing them together and properly integrating their performance becomes a separate task. On the other hand, an increase in the scale of the object may create serious communication problems and require us to rethink systemwide decision-making processes. To summarize, by stretching and compressing the scale of familiar objects and processes, and by focusing on functionality rather than implementation, we can shake off professional stereotypes and limitations. As a result, we can discover unexplored opportunities for innovation.

Playing with Time and Money

Similar to the scale tool, we can use time and money to explore constraints related to today's normal solutions. We can play with the time parameter by changing the lifetime for the object or processes of your choice, ranging from zero to eternity. Let's assume that today's laptop computer serves its user well for approximately three years. What if you had to use it for thirty years instead? Would the implementation have to be different for a physical versus a virtual laptop? For example, your virtual laptop, that is, all the applications and interfaces you have today, would have to be reproduced on any physical device during all these thirty years. On the other hand, if your physical computer would have to last thirty years, the design principles, including interfaces, reliability, upgradability, and accommodation of user preferences would have to be different as well. The durability problem often arises in relation to long-lasting physical equipment, such as nuclear power plants, space stations, submarines, airplanes, energy grids, and others. Also, if you were to design a planet-computer, which was supposed to last forever, it would have to be able to adapt to changes by renewing itself on a constant basis,* or, alternatively, to slow down all of the environmental processes so that the change would become unnecessary.

On the opposite side of the time scale, a computer lasting only a nanosecond would either have to be extremely fast or process and communicate only tiny amounts of information. Additionally, we can discover new kinds of life cycles by switching our thinking from physical to functional. For example, a regular shopping cart has a fixed design and is supposed to last for years, while a virtual, software-based one can be personalized and remain in existence for the duration of the shopping trip only. Alternatively, it can last you a lifetime, as goods and services you buy and trade pass by, forming a timeline amenable to, for example, preferences analysis. In short, speeding up or slowing down familiar processes and transactions within the imaginary computer helps us identify potential use cases and innovation opportunities.

* Similarly, an economy (or company, or university) can be thought of as a system that is supposed to last forever. From this perspective, investing in innovation becomes a matter of survival rather than choice. Such a long-term view makes obvious the transitory nature of present-day constraints and the trade-offs associated with them. In consequence, teaching people that "everything is a trade-off" turns into a recipe for a large-scale disaster.

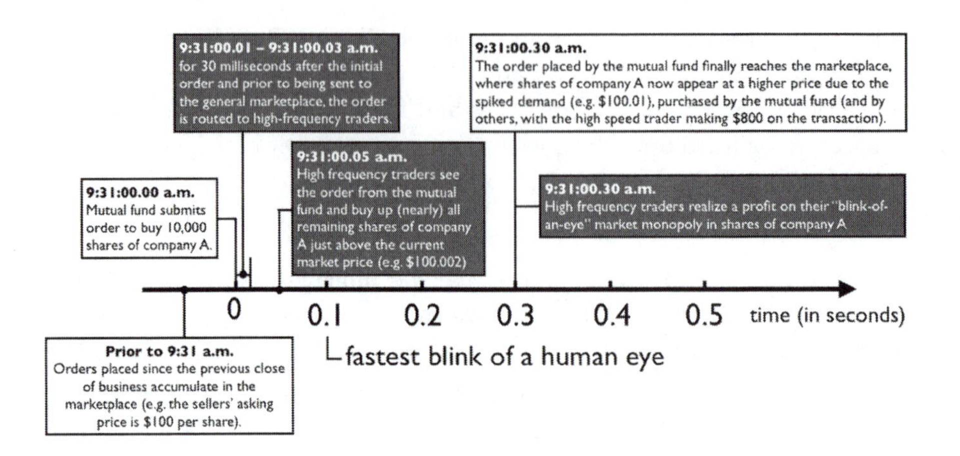

FIGURE 10.4 High-frequency stock trading process. (From Charles Duhigg, Stock traders find speed pays, in milliseconds, *New York Times*, July 23, 2009, http://www.nytimes.com/2009/07/24/business/24trading.html [29].

For example, modern stock-trading systems can execute transactions faster than any human. Whether the price of the stock is going up or down, by inserting itself between the buyer and the seller, high-frequency trading software is practically guaranteed to make money (Figure 10.4). For Wall Street firms, a few milliseconds' delay in transmitting an electronic order means a lost moneymaking opportunity. Therefore, computers that execute sophisticated trading algorithms not only have to process information fast, but also be physically close to data centers that execute financial transactions. Otherwise, even signals moving at a speed close to the speed of light would arrive too late for a successful high-frequency trade to occur.

One more way to use the time tool is to ask ourselves, "How would we solve the problem if we had a hundred years, or ten years, or one year, or one month, or a couple of days, and so on down to a nanosecond or less?" As we consider the changes in time allowed, the nature of the problem that we have to solve changes as well. Often, multiple trade-offs between short-term versus long-term thinking become more explicit during this brief exercise.

For example, 6 months is a long time for a startup strapped for cash. On the other hand, a 6-month time horizon is extremely short when thinking about the start-up's future patents. Because in the United States a typical patent can be enforced for twenty years from the filing date, many resource limitations assumed in the current short-term solution are likely to disappear in the future. The fact that the startup's technology is successful today doesn't necessarily mean that the patent intended to protect it will have any commercial value a few years from now. Even a brief discussion of short- versus long-term aspects of the problem situation helps inventors augment the hacker-style, immediate and intuitive thinking with a more deliberate, system-oriented approach that may improve the longevity of the solutions.

Similarly, we can use the money tool to play with the amount of resources available for creating solutions. A quick walk through a range of budget options, from having unlimited money to no money at all, helps bring into focus ideas that do not require significant funding for achieving desired results. On the other hand, with a lot of extra money to spend, it becomes possible to address issues requiring a massive infrastructure overhaul.

For example, today's American electrical power grid was created for the needs of the industrial age, that is, powering large-scale factory operations. Now, when a significant amount of work is done in homes and offices, producing large amounts of electricity in a handful of massive power plants and moving the electricity over long distances to a slightly larger handful of high-demand locations may no longer be needed. Except for data centers, the demand for energy in the information age might be better served by more distributed energy production from renewable resources, provided we deploy sufficient storage to smooth out demand variations. The money tool helps make such resource dependencies explicit and lets us focus on the roadmap-related aspects of the problem situation.

Imagination Development: Conclusions about the Scale-Time-Money Tool

The STM tool is a quick exercise—no more than 10 to 15 minutes—that can be used to warm up and stretch our mental "muscles" prior to an invention session. Its purpose is to challenge the normal point of view and help participants to reevaluate existing solutions. It also makes it easier for inventors and entrepreneurs to uncover hidden assumptions and constraints. During invention development, Agile Imagineering, and patent strategy sessions, the STM operator is useful for exploring the application of the original idea at several scale and resource levels, well beyond the inventor's initial intent.

REVIEW QUESTIONS

According to *Bloomberg Businessweek* (October 24, 2012): "The U.S. Transportation Department in August started a field test of almost 3,000 so-called connected vehicles in Ann Arbor, Michigan. The cars are equipped with wireless devices that use global positioning systems to communicate with other vehicles and roadside systems including at intersections" [30].

Use the STM tool set to explore major constraints and potential breakthroughs on the path to large-scale innovation in self-driving vehicles. In your opinion, will the change first affect commercial or passenger vehicles?

What impact on the infrastructure and the environment should we anticipate in five, ten, twenty, or fifty years? What trade-offs should we break to achieve mass-market adoption of the technology? If you were responsible for building a mass transit system, would you spend your budget on the infrastructure for self-driving vehicles or high-speed railways? Explain your choice.

SECTION II SUMMARY

In Section II we encountered a happy, clear-sighted turkey, whose unlucky destiny was sealed by a wrong perspective. We introduced tools for developing divergent and convergent thinking that promote multilevel thought experiments in system navigation, and could help one to avoid repeating the turkey's experience.

To overcome the barriers created by the WYSIATI (what you see is all there is) bias, we learned how to create new perspectives and stretch-test the normal reality in various dimensions: space, scale, time, and resources. Using Daniel Kahneman's formula *Great Success = A Little Bit More Talent + A Lot of Luck*, we examined inventor stories and saw how luck can be lost and found.

In combination with the system model introduced in Section I, our new flexible thinking tools will help us take advantage of system evolution patterns. In Section III we will learn how systems evolve along the S curve, creating space for invention and innovation.

REFERENCES SECTION II

1. Gell-Mann, Murray, On getting creative ideas, *Google Tech Talks*, March 14, 2007.
2. Lehrer, Jonah, *Imagine: How Creativity Works* (Boston: Houghton Mifflin Harcourt, 2012).
3. Förster, Jens, *The Unconscious City: How Expectancies About Creative Milieus Influence Creative Performance Knowledge and Space*, 2009, Volume 2, 219–233, doi: 10.1007/978-1-4020-9877-2_12.
4. Ritchey, Tom, *Wicked Problems: Structuring Social Messes with Morphological Analysis*, Swedish Morphological Society, last revised November 7, 2007, http://www.swemorph.com/pdf/wp.pdf.
5. Gelman, Eric, Fear of a black swan, *CNNMoney*, April 3, 2008, http://money.cnn.com/2008/03/31/news/economy/gelman_taleb.fortune/index.htm.
6. Arrow, Kenneth J., I know a hawk from a handsaw, in *Eminent Economists: Their Life and Philosophies*, M. Szenberg, Ed. (Cambridge and New York: Cambridge University Press, 1992).
7. Russell, Bertrand, *The Problems of Philosophy*, 1912.

8. Hume, David, *An Enquiry Concerning Human Understanding*, 1748.

9. Troianovski, Anton and Sven Grundberg, Nokia's bad call on smartphones, *Wall Street Journal*, July 18, 2011, http://online.wsj.com/article/SB1000142405 2702304388004577531002591315494.html#

10. Schewe, Phillip F., (Contribution by). *Grid: A Journey through the Heart of Our Electrified World*. (Chicago).

11. DeLong, Bradford, *The Roaring Twenties*. http://web.archive.org/web 20120210005024/ http://econ161.berkeley.edu/TCEH/SLOUCH_roaring13.html

12. Kahneman, Daniel, *Thinking Fast and Slow* (New York: Farrar, Straus, and Giroux, 2011).

13. Godin, Seth, How to get your ideas to spread, *TED Talk*, February, 2003, http://www.ted.com/talks/lang/en/seth_godin_on_sliced_bread.html

14. MIT Inventor Archive: Otto Rohwedder, MIT, March 2007, http://web.mit.edu/invent/iow/rohwedder.html.

15. Wrangham, Richard, *Catching Fire: How Cooking Made Us Human* (New York: Basic Books, 2006).

16. Murashkovska, I.N., and N.P. Valums, Introduction: Divide-Connect and Fall Back-Spring Ahead, in *Kartinka bez Zapinki*, 1995. http://www.pedlib.ru/Books/3/0286/3_0286-6.shtml#book_page_top

17. Cowan, Nelson, The magical number 4 in short-term memory: A reconsideration of mental storage capacity, *Nature* 428 (April 15, 2004): 751–754, doi:10.1038/nature02466; *Behavioral and Brain Sciences*, Cambridge Journals Online (2001): 0140-525X.

18. Feigenson, L. and Justin Halberda, Conceptual knowledge increases infants' memory capacity, *Proceedings of the National Academy of Sciences* 105, 29 (2008): 9926–9930.

19. Spence, Jonathan D., *The Memory Palace of Matteo Ricci* (New York: Penguin Books, 1985).

20. Rushdie, Salman, *The Enchantress of Florence* (New York: Random House, 2008).

21. Altshuller, G.S., Frontcover *Naitl Ideyu* (Nauka, 1980).

22. Mann, Charles C. *1493: Uncovering the New World Columbus Created* (New York: Vintage Books, 2011).

23. Levitt, Steven D. and Stephen J. Dubner, *SuperFreakonomics* (New York: William Morrow, 2009).

24. Amabile, Teresa M., Sigal G. Barsade, Jennifer S. Mueller, and Barry M. Staw, Affect and creativity at work, *Administrative Science Quarterly* 50, 3 (2005): 367–403, doi: 10.2189/asqu.2005.50.3.367.

25. Rusli, Evelyn M., Amazon.com to acquire manufacturer of robotics, *New York Times*, March 19, 2012, http://dealbook.nytimes.com/2012/03/19/amazon-com-buys-kiva-systems-for-775-million/.

26. Mick Mountz: The hidden world of box packing, Ted Talk, 2011. http://www.ted.com/talks/mick_mountz_the_hidden_world_of_box_packing.html
27. McClelland, Mac, I was a warehouse wage slave, *Mother Jones*, March/April 2012, http://www.motherjones.com/politics/2012/02/mac-mcclelland-free-online-shipping-warehouses-labor?page=2.
28. Petrov, V.S. Operator RVS. 2002. http://www.triz.natm.ru/articles/petrov/7.1.2.htm
29. Duhigg, Charles, Stock traders find speed pays, in milliseconds, *New York Times*, July 23, 2009, http://www.nytimes.com/2009/07/24/business/24trading.html?_r=1.
30. Keane, Angela Greiling, Driverless car future sends Google, U.S. to figure rules, *Bloomberg News*, October 24, 2012, http://www.businessweek.com/news/2012-10-23/google-carmakers-discuss-self-driving-cars-with-u-dot-s-dot-regulators.

Section III

System Evolution and Innovation Timing

A meaningful simplest case stimulates the will to think by reducing the threat of being forced to accomplish repugnant and tedious tasks. [1]

—**William Shockley***

* Co-inventor of the transistor; winner of the 1956 Nobel Prize in Physics.

11 The S Curve
Dynamics of Systems Thinking

In his 1942 book *Capitalism, Socialism, and Democracy*, economist Joseph Schumpeter coined a new term, *creative destruction*, to describe the process "of industrial mutation ... that incessantly revolutionizes the economic structure from within, incessantly destroying the old one, incessantly creating a new one." Today, seventy years after Schumpeter introduced the concept, creative destruction is widely considered to be essential to sustained economic growth [2].

Entrepreneurs drive the creative aspect of the process. The seek new opportunities, create new markets, and along the way, destroy old business models. For example, Edison's electricity distribution system drove gas lights into extinction. Automobiles and tractors eliminated horses from streets and farmland. Electric refrigerators made ice making largely obsolete. Electronic books (e-books) will have made paper books belong to museums, and so on. We take for granted that technologies, while consistently bringing us free lunch, come and go. The primary purpose of this chapter is to show the connection between innovation-driven growth and patterns of system evolution. To help determine the right timing for introducing creative solutions, we will explore in detail when, how, and why problems emerge in the system.

At the aggregate level, the process of innovation is often described by an S curve (see Figure 11.1 a, b, c, and d). The curve exhibits several distinct periods of evolution. In the beginning, it undergoes an incubation period, when a lot of creative effort does not seem to produce tangible performance improvements. Then, after reaching an inflection point, the system undergoes a period of rapid growth, a pattern often called the *hockey stick* due to the nonlinear character of the transition. Eventually, the growth slows down, the system enters stagnation or even decline, and is replaced by a new one that can be described by another S curve.

Market adoption charts frequently show the process of change going through several key stages: first, the new product or service is embraced by innovators, then picked up by early adopters, who pass it on to early majority, later joined by late majority, laggards, and finally abandoned by everybody in favor of a new product or service (Figure 11.2).

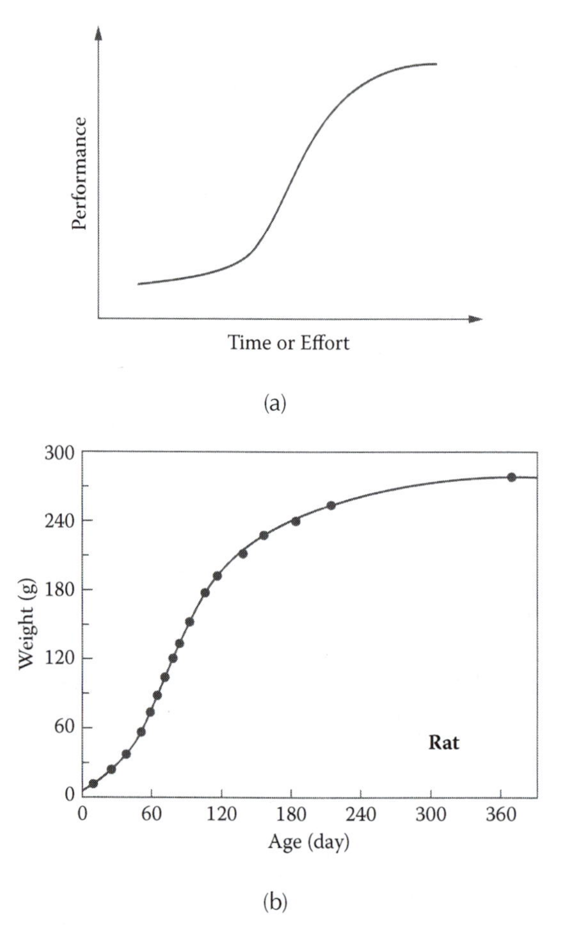

FIGURE 11.1 (a) A generic S curve: Time/effort-performance. (From R.A. Burgelman, C.M. Christensen et al., *Strategic Management of Technology and Innovation*, 2009 [3].) (b) A typical age-weight S curve for a mammal. (From Geoffrey West, Ted Talk, July 2011 [4].) (continued)

In the 1980s, very few people used email because the Internet was confined to major universities and some government institutions. Later, proprietary networks like MCI, America On-Line (AOL) and Compuserve began providing email services to their subscribers. Proud of their digital roots, some early adopters of the technology still keep their old @aol addresses as badges of technological prowess. As the web took off in the mid-1990s, email became free for the early majority of users, with services like Rocketmail and Hotmail offering access to email via browser-based interfaces. Eventually, after IT giants such as Yahoo, Microsoft, and Google entered the market for unlimited-storage email services, almost everybody in the world, including members of the very late majority, got at least one email account. Even governments, which for centuries relied exclusively on paperwork, turned

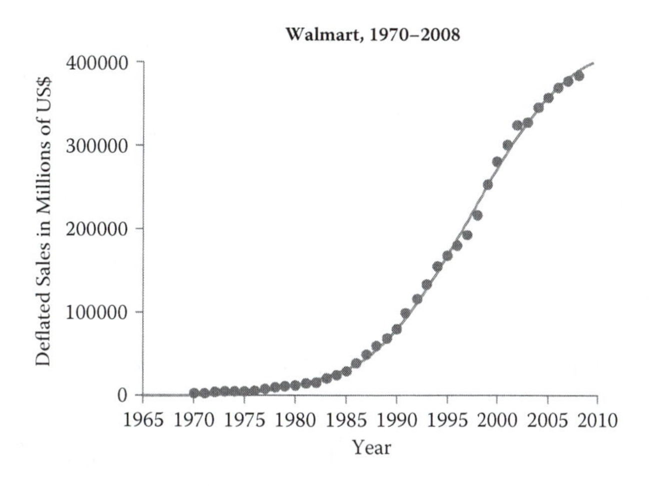

(c)

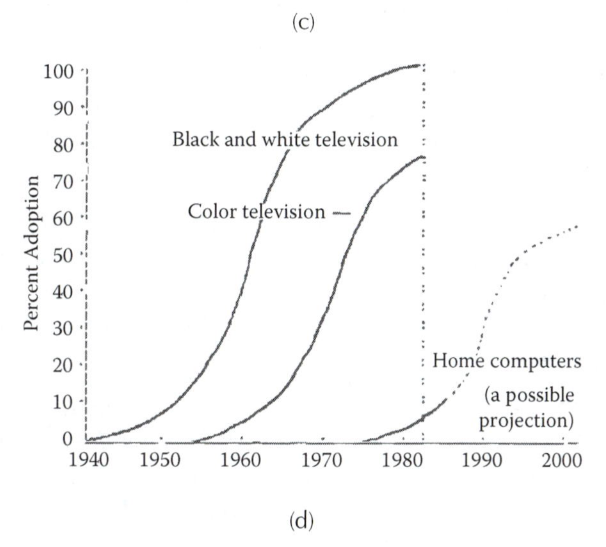

(d)

FIGURE 11.1 (c) Walmart: Time-Sales S curve. (From Geoffrey West, Ted Talk, July 2011 [4].) (d) Market adoption S curve. TVs and PCs. A 1984 actual and forecast. (From Everett M. Rogers, The diffusion of home computers among households in Silicon Valley, *Marriage & Family Review* 8, 1985 [5].)

to digital communications. In October 2010, Sergey Sobyanin, then newly appointed mayor of Moscow, Russia, shocked his long-serving city bureaucrats by requiring that all of them be capable of sending and receiving emails [6].

We can also see that electronic communications, either functionally equivalent or better than email, migrate to social media services, such as Facebook, Twitter, LinkedIn, and others. Email is a long way from dying, but it has become just one of many communication streams entering our

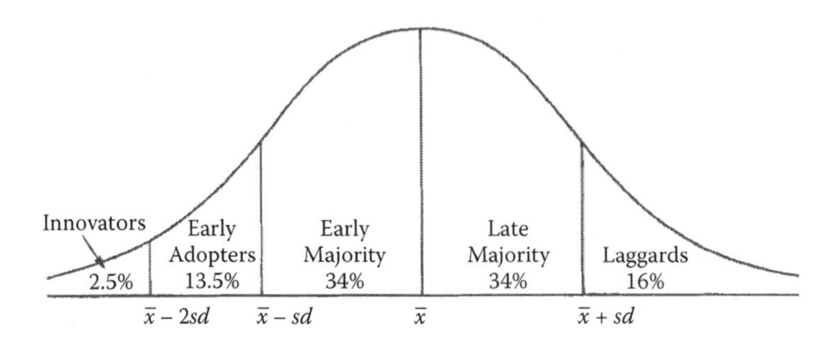

FIGURE 11.2 Market adopter categorization. (From E. M. Rogers, *Diffusion of Innovations*, 5th ed. 2003 [7].)

information spaces. As a technology, it is innovation-poor, unlike it was during the decade of the 1990s when it grew rapidly in features and the scale of user adoption. The latest round of minor improvements occurred in the 2000s, spurred by the introduction of smart phones, such as Blackberry and iPhone. We can reasonably expect a drastic reduction in email-specific innovation during the 2010–2020 decade. This transition can be reflected on the S curve as the last, slowdown segment.

Typically, S curves are produced for light bulbs, radios, audio tape players, TVs, personal computers, phones, and other devices, which in system terms represent instances of the Tool (see Figure 11.1d). The overall S-curve pattern is quite stable, and we can rely on it for exploring the timing of innovation-related changes within the system.

In earlier chapters we saw how to model a system using five functional elements: Tool, Source, Distribution, Packaged Payload, and Control. Now we are going to combine the concept of the S curve and the system approach to identify problems and discover innovation opportunities. First, we will consider what happens to system elements during the initial phase of the S curve.

12 A Stage of System Evolution

Synthesis

The class of problems typically encountered at the foot of the S curve we will call *Synthesis*. During this stage of innovation, all the functional elements have to be put together, so that the system as a whole can be capable of providing its new functionality at scale. Before that happens, the chances for user adoption are very low. For example, prior to Edison, the first electric lighting systems were less reliable and had less utility than the established gas lighting infrastructure. Light bulbs would burn out unexpectedly, and being connected in series, break the flow of current, plunging entire buildings into darkness. Wires could be easily broken and, due to poor insulation, would cause fires or inflict severe electric shocks on workers and bystanders. In addition to that, steam engines connected to dynamos would produce noise, smoke, and sparks right in the middle of town. Sometimes they would blow up or let off torrents of steam, making residents concerned about the safety of the entire enterprise. Under the circumstances, to replace the good old gas lighting with the newfangled electric light bulbs would be an act of madness.

Edison's great achievement was synthesizing an electric system, which as a whole worked much better than its gas alternative. All functional elements were improved to such a degree that by working together they achieved a new level of performance. For the early adopters, jumping onto the new technology curve was still a bit risky, but no longer suicidal.

Every element in the new electric lighting performed better than its functional equivalent in the gas system. The light bulb was brighter and safer than a gas burner emanating open fire. The improved dynamo and steam engine were better than gas storage tanks that occasionally would leak highly flammable materials. Insulated electric cables buried underground were invisible and, more importantly, reliable. The new switches could control multiple electric lights with just one flick. They were superior to the drawn-out process of lighting gas lamps, that is, one-by-one, using open-flame torches.

Edison was quick to show off to the public this particular advantage of his electricity distribution system. Being aware of the media's need for exciting news—newspaper circulation was expanding—Edison invited reporters to visit his invention lab in Menlo Park, New Jersey. When they approached the lab, he turned on electric switches, instantaneously lighting up the path. This sudden burst of light made a powerful impression on reporters, who were, like everybody else at the time, used to gradual firing up of gas lamps. The next day, all the major newspapers carried stories about this and other incredible inventions by Edison.

The inventor's technical and entrepreneurial efforts, backed by J. P. Morgan's venture capital, set the stage for the creative destruction of gas lighting, which was the dominant lighting system of the second half of the nineteenth century. Edison's system worked. It scaled to at least the level of gas lighting, was safer, and more controllable. But the further need for scale emerged only with the growth of cities and industries in America. In places like London, where gas lighting was an entrenched system, conversion to electric lights occurred decades after Edison's first successful Synthesis in Manhattan. It took a genius to not only put all of the technological elements together, but also convince the public to invest in and adopt the new system as a whole.

Solving Synthesis problems is usually a long-term proposition, because they require extended experimentation with multiple technological solutions. Since stable interfaces between different system elements are yet to be developed, independent, parallel creation of reliable instances of the elements is rather difficult. Synthesis is slow, but once it happens, the process of creative destruction becomes inevitable because customers begin moving into the new world created by breakthrough innovation.

THE DESTRUCTION OF PAPER BOOKS: EVOLUTION OF A SYSTEM

Let us consider a major technology revolution that is under way in large-scale consumer information systems: the transition from paper books to electronic ones. Note that printed books have been around for about half of a millennium, since the time of Gutenberg. The form of the book itself goes back even further, to the invention of the codex, a collection of numbered pages bound together between two protective covers. Since that time, the book and related print media have become an essential part of a large variety of systems and routine processes. Most of those systems will be destroyed right before our eyes within a relatively short period of time.

Attempts to create electronic books are not new. For example, in 1979 David Rubincam of West Hyattsville, Maryland, patented an e-book with holographic memory as its main component [8]. The interfaces of the device looked remarkably similar to the first Amazon Kindle (Figure 12.1): a display

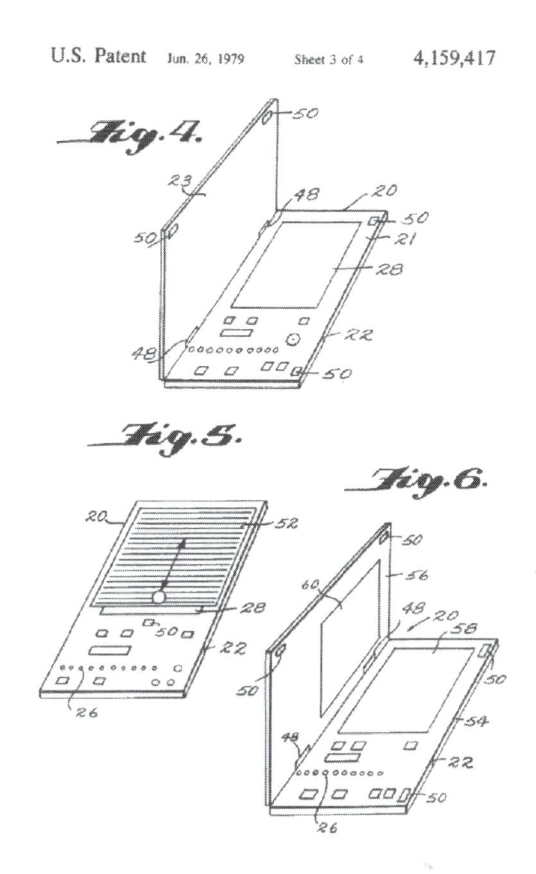

FIGURE 12.1 Electronic book, US Patent 4,159,417. David P. Rubicam [8].

for text, buttons to navigate between pages, and a connection for downloading the text from a computer. Nevertheless, it took more than twenty-five years before the concept of a ubiquitous e-book started taking off. Why?

An easy place to start answering this question would be the e-book reader, which in system terms is an instance of the Tool. A modern e-reader, for example Kindle or iPad, is now at least as good, if not better than the traditional paper book in terms of readability and portability. With display image densities from 200 to 300 pixels per inch, the e-reader is easy on human eyes. Depending on the specific display technology, we can read electronically both in the broad daylight of the beach and in the total darkness of the bedroom at night. Moreover, the device holds hundreds, if not thousands of texts, making it great for travel. It fits easily into a pocket or a small purse. Most electronic texts are searchable and their fonts are adjustable. A connected e-book allows us to share comments, see notes and highlights other readers left in the text—a very important feature for group study or nonfiction reading. Of course, the device is far from perfect, but in the pure readability department, it is on par with the paper book. Additionally, the e-reader has valuable features that are simply unavailable through the traditional paper-based technology.

Although the price of the e-reader is over ten times higher than that of the paperback, when you take into account that it potentially holds not just one book, but rather an entire library, the difference in price for a person who reads regularly becomes negligible. From a system perspective, the e-reader is an instance of the Tool. It delivers the system's key functionality—convenient access to e-books and related information. Moreover, manufacturers can produce e-readers in huge volumes at reasonable prices. Although Rubincam's 1979 invention did not fulfill these basic requirements, it would be a mistake to attribute the failure of the hologram-based device exclusively to the poor quality of its memory or display. Rather, it was early for its time not only because of its own technical flaws, but also because the rest of the system was simply missing.

It is a recurring pattern in the history of innovation that new ideas are frequently either overlooked or overhyped. As we discussed earlier in this book, when in the late 1990s Stanford grad students Larry Page and Sergey Brin, the future founders of Google, tried to sell their relevancy-based search technology to established Internet companies, their offer was met with apprehension. For the potential buyers, the $1.6 million price tag of the solution seemed too high. Just a few years later, Google put the very same companies out of business, using the rejected technology to become a dominant force in the industry.

> The Stanford product was too good. If Excite were to host a search engine that instantly gave people information they sought, he explained, the users would leave the site instantly. Since his ad revenue came from people staying on the site—"stickiness" was the most desired metric in websites at the time—using BackRub's technology would be counterproductive. [9, p. 30]

By contrast, another innovation created in Silicon Valley was met with much greater expectations. In 2001, the famous inventor Dean Kamen, after working on the concept for over a decade, created a high-tech personal transporter, Segway, (code-named "Ginger"). On December 2, 2001, *Time* magazine wrote that the device got "over-the-top assessments from some of Silicon Valley's mightiest kingpins. As big a deal as the PC, said Steve Jobs; maybe bigger than the Internet, said John Doerr, the venture capitalist behind Netscape, Amazon.com and now Ginger." It was even predicted that "the Segway Co. will be the fastest outfit in history to reach $1 billion in sales" [10]. Despite all the high-profile endorsements, Segway never made the impact it was supposed to produce.

As we mentioned earlier, it is natural for people to experience the *what you see is all there is* (WYSIATI) effect, that is, focus their attention on a specific solution, forgetting about the rest of the system. A check for system completeness is the first step in the process of escaping the tendency to overhype or overlook a new idea based on features that seem most salient at the moment.

In the e-book example, when we turn our attention to the Source, we can see that now, unlike thirty years ago, there are libraries with millions of electronic texts available on the Internet. Moreover, modern texts are created electronically from the early stages of the book publishing process. Therefore, no book scanner is necessary for editing work or putting a particular text online. Due to the rapid development of graphics software, illustrations can be produced and embedded into the text electronically. Not printing a book, not having the need to store it in a physical warehouse or on a bookshelf provides a cost advantage for book retailers. Unlike paper books or holographic memory hardware, a copy of an e-book can be manufactured instantaneously, at near-zero cost. Adding pictures, colors, sounds, and other aspects traditionally reserved for high-end publications no longer entails a dramatic increase in production costs. The author is no longer caught in the trade-off between the need to save paper and the desire to include as many figures as necessary to explain or demonstrate his or her idea. To summarize, with new Source implementations, book copies can be stored and generated at extremely low costs.

With the proliferation of 3G and LTE wireless networks, Distribution has finally caught up with book lovers' desires for instant gratification. Unlike the case with the early Internet shopping for physical books, there is no multiday delay between book purchase and experience. There is no need to go to the store and stand in line either. We can get our books practically anywhere at any time. The further wireless networks reach and the more wireless bandwidth is available, the more media-intense books we will have ready for immediate delivery. With a direct connection to a wireless network, the e-reader doesn't need a traditional personal computer (PC) as an intermediary for content download and transfer. On the system completeness checklist we can confidently check off the Distribution.

As an instance of the Packaged Payload, digital text has multiple advantages over its analog peers; it's easy to store, transfer, index, and adjust according to user reading preferences. While the earlier digital text formats, such as PDF and Postscript, were originally developed for printing, today's formats are being created for viewing on a screen, taking advantage of new device interfaces and display capabilities. For example, even basic modern digital book formats allow authors to change fonts or include color illustrations, which drastically improves the author's ability to show and explain without a significant increase in production costs. The *show* aspect is particularly important for textbooks, children's books, research papers, and other publications, where one picture is worth a thousand words. Moreover, the content can be made interactive. Given the social nature of the learning process, the e-book enables immediate feedback on the quality of its reader's experience. With slight modifications, digital texts can be experienced on a wide variety of reading devices without any loss of quality. In short, instances of the Packaged Payload in the system are already better than their analog competitors, and the technology shows a significant potential for further growth.

Finally, the Control element is also performing well. For example, Amazon recommendations and reviews that facilitate the process of e-book selection are the same as for paper books, an important shared system element that reduces psychological barriers to user adoption. In the event that a user finds a paper book to buy, a less expensive e-book version is usually offered for purchase. Another advantage is that the user can choose the specific device or even multiple devices on which to download the e-books. As more people switch to digital books, many physical libraries, both public and private, are being converted to provide a wide variety of services to consumers. Buying, renting, subscribing, borrowing, and other business models can be easily made available in the future. Along with other instances of the system's elements, the Control works to the advantage of e-books. Furthermore, all elements show good potential for growth.

As we can see, the Synthesis problem in the e-book space has been solved. All elements of the system are functional and some of them provide remarkable advantages over their predecessors in older systems. This was not the case twenty-five or even five years ago, while various industry players worked on the specific implementations for different system elements. At the time of writing of this (e-)book, the current system position on the S curve for distribution of electronic texts is just above the inflection point. On the other hand, the S curve for good-old paper books entered the decline phase. As users leave this system behind, its infrastructure is likely to deteriorate.

CONCLUSIONS REGARDING THE SYNTHESIS STAGE OF A SYSTEM

The initial flat portion of the S curve corresponds to an extended effort necessary for solving a Synthesis problem. Not one, not two, but *all* elements of the system have to be instantiated and interfaced with each other for the system to succeed. Understanding the nature of the Synthesis problem helps us appreciate the time, resources, and innovation efforts necessary for creating new systems. A typical mistake at this stage is to hype the emergence of a viable implementation technology for one particular element, most often either Tool or Source, and forget about the rest of the system.

For example, before Apple's iPod, MP3 audio players like Diamond Rio, Creative Jukebox, and others received a lot of attention from the media and got significant support from digital media enthusiasts. Despite being first to market, they failed to capitalize on the advantage. This happened not only because their user interface was less refined than that of the iPod. From a system perspective, all of them lacked at least a functional Control component necessary for intuitive management of large music collections and content synchronization. As an instance of Control, Apple's iTunes was vastly superior to its peers—Music Match and Microsoft's Windows Media Player, software programs originally developed as audio/video playback Tools for the PC. In addition to integrating iTunes with the iPod and providing an interface for fast transfer (i.e., the Firewire cable), Apple developed podcast formats, custom playlists, media libraries, and software tools for media creation. All this made it easy for consumers to feed their players with vast amounts of content.

Similarly, Amazon's Kindle has a strong market advantage over other electronic book readers not because of its superior hardware features, but because of Amazon's great ability to facilitate selection, purchase, and delivery of electronic books. From a system perspective, we can see that the real competition in the e-book space is unfolding between Amazon and Apple, the latter company in control of iPad devices and cloud-based iTunes content services.

Though an exceptional performance from a specific element (the light bulb, the e-book reader, the iPod) can be critical for the creation of the wow factor that attracts consumer attention, in the Synthesis stage it is not sufficient to sustain the initial success. Even worse, the breakthrough creates a lot of hype and soon leads to a market disappointment (Figure 12.2). In the meantime, the advance in technology attracts other industry players. They learn from the original innovator's mistakes and become successful by putting together the whole system, either alone or as a coalition. Though the newly synthesized system works reliably in a relatively limited set of scenarios, its initial success sets the stage for *early growth*, the next phase of development along the S curve.

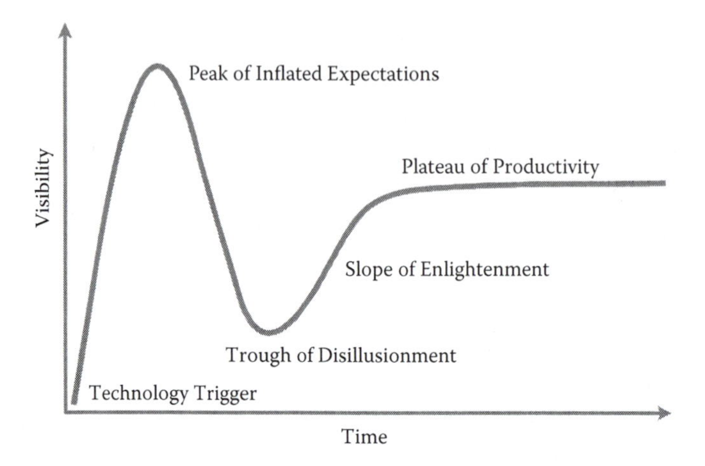

FIGURE 12.2 Gartner hype cycle. (From http://en.wikipedia.org/wiki/Hype_cycle.)

REVIEW QUESTIONS

1. The chart below shows the projected spread of the worldwide obesity epidemic among 45- to 59-year-olds in different parts of the world. Identify at least one system responsible for the epidemic. Where on the S curve does it belong? What system would you synthesize to address the problem? Sketch out its key elements.

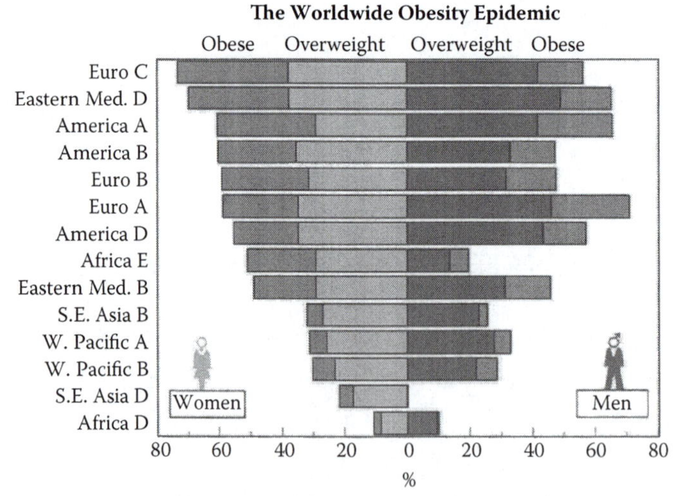

Source: Dr. Philips T. James et al., The worldwide obesity epidemic, *Obesity Research* 9, 2012 [11].

2. According to *Nature* [12]:

   > Researchers in Germany have claimed a breakthrough: a material that can act as a superconductor—transmit electricity with zero resistance—at room temperature and above. Superconductors offer huge potential energy savings, but until now have worked only at temperatures of lower than about –110°C.

   Assuming that the researchers succeed in creating a practical material suitable for all industry applications, what systems would you synthesize? Sketch out systems that transfer and transform mass, energy, and information. Use the Scale-Time-Money (STM) tool to come up with ideas for macro and micro systems.

13 A Stage of System Evolution

Early Growth

After the Synthesis problem is solved, the system begins to grow along two paths: addition of instances of elements, most often Tools, Sources, and Packaged Payloads, and replication of the system as a whole. For instance, Edison's system grew rapidly through the increased demand for light bulbs (Tools) and generators (Sources) that fed the bulbs with electricity within the same power distribution area. As a result, power (Packaged Payload) generation and consumption grew as well. Additionally, Edison's company built and installed complete electricity systems in places where coal could be delivered easily.

Similarly, once Amazon put the e-book distribution in place, the company aggressively began promoting sales of its Kindle devices and made deals with publishers to create an increasing supply of inexpensive digital content. To pump out even more content, Amazon started its own self-publishing program for aspiring authors who couldn't get attention from the more traditional publishing houses. In addition to physical e-reader devices, Amazon developed software e-book readers for smartphones and tablets, compatible with its e-book formats. As a result, the sales of Kindle devices and e-books grew rapidly. By mid-2010, e-books were outselling hardcover books by eighty percent despite the fact that there were millions of hardcovers compared to 630,000 Kindle book titles. According to the *New York Times* [13]:

> The shift at Amazon is "astonishing when you consider that we've been selling hardcover books for 15 years, and Kindle books for 33 months," the chief executive, Jeffrey P. Bezos, said in a statement.

The competition tried to replicate (albeit much less successfully) the Amazon book system. For example, Barnes and Noble developed a structure very similar to that of Amazon's, with digital books, an online store, customer recommendations, Nook e-reader, and so on. Sony and Apple got into the e-book game as well, though Apple developed an iPhone/iPad software reader only.

In the Early Growth stage, which is the main focus of this chapter, the innovator's ability to ramp up the system's output is essential for success. In the example of the bread-slicing machine we considered earlier, after the 1929 stock market crash, Otto Rohwedder, the original inventor, lacked the capital necessary to deploy his solution on a large scale. In addition to that, his patents were not strong enough to stave off the Wonder Bread invasion. But before we discuss the role of patents and industry structure during Early Growth, let us first introduce the concept of *dominant design*. It is a key concept necessary for recognizing opportunities and perils facing innovators who are striving to move up the S curve.

SIGNS OF EVOLUTION: THE EMERGENCE OF A DOMINANT DESIGN

The most important innovation outcome of the Early Growth phase is the emergence of dominant designs. Dominant design is an implementation of a new functionality that the market adopts as the prototype for future implementations. Simply put, the dominant design is what everybody thinks all products or services in the new category should imitate. One glance at something made according to the dominant design invokes an instant recognition of what to do with it (see, for example the 2,000-year-old dominant design for a bathtub in Figure 13.1).

In another example, since Edison's time more than one century ago, we all know what a light bulb looks like. Recalling an earlier interface example, all keyboards, even the virtual ones for smartphones, require the QWERTY layout, a design once created for mechanical typewriters (Figure 4.4). In a more recent example, after trying for years to come up with an interface for a social networking service of its own, Google gave up and largely copied the layout of Facebook pages [14]. The innumerable copycat implementations of iPhone-like user interfaces tell us that Apple has created a dominant design for the new generation of smartphones. The same applies to tablets and Apple stores. In 2011, a journalist traveling in China discovered that some local entrepreneurs created exact copies of Apple stores, including employee uniforms, company logo, and service standards [15]. Dominant designs are difficult to invent, but easy to steal.

Early growth is about scaling up, that is, adding to the system replicas and variations on the main theme. The simplest way to accomplish this is to rip off someone else's successful idea. Imitation is traditionally considered to be the sincerest form of flattery; often, this is accompanied by very

FIGURE 13.1 Dominant design: An ancient Roman bathtub in the Science Museum in London. The vast majority of modern bathtubs have the same form and function as this 2,000-year-old exhibit.

unpleasant consequences for the inventors. For example, in 1793, a young Yale-educated lawyer named Eli Whitney, who moved to the American South in search of job opportunities, invented the cotton gin. By dramatically improving the speed of processing raw cotton, the invention created a textile revolution. In a letter to his father Whitney wrote, "One man and a horse will do more than fifty men with the old machines. ... Tis generally said by those who know anything about it, that I shall make a Fortune by it" [16].

Whitney's hopes were not realized. Though he applied for and received one of the first US patents, copycat manufacturers ignored his intellectual property rights and built their own versions of the device. After protracted legal battles, when the government finally decided to enforce the law, there was only one year left before the expiration of the patent. Whitney's invention dramatically improved productivity in the textile manufacturing industry worldwide, but financially, the imitators ripped off the inventor.

Usually, a dominant design emerges [17] after a prolonged period of experimentation during the Synthesis phase. The design is not necessarily the most novel, efficient, or beautiful. Instead, its main advantage is how well it fits into the rest of the system, while delivering a new system functionality at scale. Paradoxically, the dominant design often comes out quite imperfect, but succeeds despite its many imperfections. For example, Edison's light bulb was rather dim and didn't last long. But it was the best light bulb that fit into the inventor's low-current DC electricity distribution system. In another example, Apple released the original iPhone with no support for multitasking, no Copy-Paste capability, a nonremovable battery, no picture messaging or forwarding, no iChat, and so on. Nevertheless, the light bulb, the iPhone, and even the Kindle, with its original clunky interface, succeeded in enabling rapid Early Growth.* As we discussed in previous chapters, the success of the dominant design is no guarantee of success for its inventor or owner. In the next section we will consider various ways of dealing with the copycats who are trying to exploit somebody else's great idea.

DEALING WITH COPYCATS DURING THE EARLY GROWTH STAGE

Depending on the business model and industry environment, to succeed during Early Growth, the innovator can either fight or encourage copycats.

> After the CES, we found out about a new computer from a company called Franklin. It supposedly looked a lot like ours. It arrived at our building, and it looked so much like the Apple II I was very interested.
>
> I thought, Hey, great. They copied my design. I wonder how much of it they copied. I didn't expect they would've copied much of it. I figured engineers are trained to invent and design their own things. An engineer would never look at another person's design and copy it, would they? No, that's what they go to school for. They go to learn how to design their own things.
>
> I walked over to the main building to look at it. There it was, and I was shocked. The printed circuit board inside was exactly the same size as ours. And every single trace and wire was the same as ours. It was like they'd taken our Apple II board and Xeroxed it. It was like they'd just Xeroxed a blank Apple II board and put in the exact same chips. This company had done something no honorable engineer would've done in their effort to make their own computer.
>
> I couldn't believe it.
>
> **—Steve Wozniak, *iWoz* [18, p. 220]**

* In another example, a July 2, 2012, *MIT Technology Review* article notes that the incredibly popular iPad is "far short of the ideal." Specifically, the author complains that "magazine apps combine the worst features of print and online reading" (http://www.technologyreview.com/view/428397/why-are-we-in-denial-about-the-flaws-of-tablets/). Despite all these real and perceived flaws, the use of tablets grows, while the readership of "ideal" printed magazines falls.

Fighting Copycats with Patents

One way to fight imitators is through the patent system. For example, Thomas Edison, being in the business of manufacturing and selling light bulbs, sued other manufacturers, including George Westinghouse, for patent infringement. A hundred years later, Apple found itself in a similar situation when electronics manufacturers copied key dominant design features of its iPod, iPhone, and iPad devices. The company sued and for a little while was able to deter the copycats.

One paradox of dominant design solutions is that after a lot of people use them for an extended period of time, the solutions begin to feel obvious. This is hindsight bias, a well-researched psychological phenomenon, which among other things, is associated with availability heuristics, a tendency to judge the probability of events by the ease with which they come to mind [19]. That is, once a person becomes familiar with an idea, it is really easy for him to overestimate the chances of the idea occurring in his own head. Since patents can be issued only for nonobvious solutions, a patent that covers dominant design features in a popular device can be invalidated in court due to the hindsight bias experienced by the judge or the jury. To counteract this tendency, it is usually a prudent practice to develop several patents that cover various aspects of dominant design features.

Since patents often issue years after the date of application, inventing for dominant design has to happen during the earlier stages of the S curve. This requires a forward-looking invention and patenting strategy developed to go hand in hand with research and development efforts. A part of the strategy can be proactive purchase or exclusive licensing of essential patents held by others. Of course, the strategy works only in jurisdictions that respect and enforce intellectual property (IP) rights. Obtaining and enforcing foreign patents is an expensive proposition; therefore, a country's IP policies have to be taken into account when deciding when and where to patent.

Keeping Secrets from Copycats

Another way to beat imitators is by keeping your inventions secret. Unlike devices that can be easily bought, disassembled, and reverse engineered by competition, some technical solutions can stay hidden from competitors for a sufficiently long time to obtain a dominant market share. For instance, Google kept its search and other algorithms to themselves, rather than sharing them with the industry. Moreover, when in 2010 hackers in China penetrated the company's servers and reportedly stole certain encryption algorithms, Google closed advanced technology development in that country and enlisted help from the American government to deter IP theft in the future.

Globalization, including technology transfer to jurisdictions with limited enforcement of patents, may require product redesign and enhanced security procedures, so that key parts or software components cannot be easily copied. Not keeping your important solutions secret can be disastrous for business. In the spring and summer of 2011, the American Superconductor Corporation (AMSC) lost three quarters of its market value when its main customer, Sinovel, a Chinese wind turbine manufacturer, refused to accept contracted shipments. In their efforts to reduce costs, Sinovel shifted production of AMSC-designed electronic components to local suppliers, leaving the American company empty-handed [20].

In September 2011, after an extensive investigation, AMSC disclosed that one of its employees in Austria had illegally transferred proprietary software code to Sinovel. Further, AMSC alleged that Sinovel had taken steps to alter the code so that Goutong, a company founded by Sinovel, could manufacture parts instead of American Superconductor. AMSC has sued, and at the time of writing this book, the legal proceedings are still ongoing [21]. Nevertheless, the damage has been done and many people have lost their design and manufacturing jobs due to the theft of technological secrets. As we can see, coming up with a good solution is just one part of the innovation process. Inventing ways for keeping the solution secret can be even more important than the solution itself.

Encouraging Copycats

Enabling growth through rampant copying can also be a great business strategy when companies are engaged in a market contest generally described as the War of Attrition. In game theory, the simplest War of Attrition is a competition between two players for a single object. In our case this would be a dominant business position in a new growing market. To attain the object, both players make investments over a period of time and the game ends when one of the players drops out or yields a significant market share. At the time of writing, Google is playing a War of Attrition against Apple by encouraging iPhone and iPad clone manufacturers to produce a wide variety of copycat products. As we discussed earlier, intense competition turns the devices (Tools in system terms) into a commodity, hurting the other player's ability to make money through investment in long-term growth.

Google not only provides its partners with the Android operating system software, but intends to give them access to a patent portfolio purchased from third parties. With the portfolio, the manufacturers can countersue

Apple in the United States and international courts. In 2011, Google bought the Motorola mobile device business for $12B, stating that it intends to use thousands of Motorola's patents to protect the Android-based smartphone ecosystem against Apple's IP attacks. Google is pouring its money, made in the Internet advertisement business, as well as its software expertise, into activities aimed to reduce Apple's market share. As we discussed in Chapter 5 on Control Points, the more hardware clone devices run Google's operating system, the greater the chances for the company to sell its services to businesses and consumers who use the devices.

During the period from the mid-1980s until the end of the 1990s, Microsoft used a similar War of Attrition strategy to compete in the PC market. In collaboration with Intel, the company encouraged manufacturers to produce a wide variety of personal computers compatible with the hardware architecture originally designed by IBM. In addition to hurting Apple, a rival PC technology developer, the strategy generated billions of dollars in licensing revenue. Apple's attempts to block Microsoft from using the windows-based graphical user interface (GUI) failed because the courts decided that Apple could only defend specific GUI features, rather than the entire look and feel of the interface. Having learned his "PC lesson," Steve Jobs eagerly created an extensive patent and trademark portfolio to cover specific design features when Apple developed the new multitouch interface for smartphones and tablets. This portfolio might have helped the company to deter or at least delay the most blatant attempts to copy their ideas.

Nevertheless, commoditizing Sources and Tools (especially the Tools) is a powerful strategy that can reduce or eliminate the first mover's advantage during the Early Growth stage. Industry players that are strong in Distribution and Control, or simply have deep pockets, may be able to scale up faster than the original inventor. In life science industries, such as pharmaceuticals and medical devices, where research and development cycles are long and solutions enjoy strong patent protection, patents can be used to deter Wars of Attrition. In other industries, such as consumer electronics, information technologies, and fashion, the speed of innovation, de facto standards, and industry alliances tend to work better than patent-only strategies against commoditization.

Licensing Inventions to Copycats

Licensing intellectual property, including patents and technological know-how, is a good strategy that can help first movers to benefit from growth. Rather than wasting money in a cutthroat market, you can make money by encouraging multiple industry players to produce and sell as much as they can.

At the end of the nineteenth century, Edison and Westinghouse spent years litigating patent rights on the dominant design of the incandescent light bulb. By contrast, in the early twentieth century, General Electric developed a standard for compatible light bulbs and licensed it to multiple manufacturers. As a result, the company and consumers benefited greatly from increased competition.

In another example, Philips and Sony created a vibrant audio CD market in the 1970s when they proactively licensed their technology to the consumer electronics industry. Unlike the VHS versus Betamax format war fought in the videocassette business, the CD technology transfer program helped the market to take off in the 1980s. The strategy helped the companies to make billions of dollars in licensing revenue. That was in addition to the revenue from devices and disks sold by both companies on the consumer market. Furthermore, rapid proliferation of audio CDs created a bandwagon effect, and eventually led to the development of PC-based CD-ROM standards and later to DVD. In all these cases, technology licensing facilitated market adoption and amply rewarded the original technology developers.

As we discussed in Chapter 5 on Control Points, Intel and Apple pursued the IP licensing approach to encourage the development of the digital accessories market. In these and other examples, from container shipping to McDonalds to Facebook, the key to successful licensing is in the development of inventions around system Control Points, for example, payloads and interfaces. Being the first mover, you have a chance to learn by doing, that is, encounter and solve problems in a brave new technology-market world. When you take a system point of view and look beyond the initial success or failure, you have the opportunity to create long-term value by sharing your solutions, rather than fighting wars of attrition with powerful copycats.

CONCLUSIONS REGARDING THE EARLY GROWTH STAGE OF A SYSTEM

In the Early Growth stage, scalability is the key problem. Since many decisions during the Synthesis portion of the S curve are based on trade-offs (e.g., due to limited knowledge of the market and multiple resource constraints), the innovator has to identify and break those trade-offs that prevent the system from scaling up. On the one hand, he wants to open up the system to speed up market adoption. On the other hand, when strong players enter the market, he runs the risk of losing control over the system's development.

A good way to address this problem is to obtain control over dominant designs for Tools, Packaged Payloads, and—importantly—interfaces between different elements of the system. The goal is to own de facto standards in the emerging market, for example, through speed, patents, business alliances, wars of attrition, or a combination of all of the above.

At this stage, the value begins to shift from the Source–Tool to Distribution–Control axis. The system continues to succeed despite the lack of supporting infrastructure. Eventually, as more and more Sources and Tools are added at a rapid pace, Distribution becomes a bottleneck. The system can no longer sustain its growth due to various infrastructure problems.

14 A Stage of System Evolution

Distribution Buildup

As we discussed in the previous chapter, the driving principle behind the Early Growth phase is "more is better." When it is not supported by massive investments in Distribution, this principle eventually turns a recipe for success into a recipe for disaster. That is, the more Tools, Sources, and Packaged Payloads are added to the system, the more congested the Distribution routes become. Initially, the problem can be solved by traffic optimization, but sooner or later a major overhaul is required. As we discussed earlier, Edison's DC-based electricity system satisfied Early Growth in lighting and small-scale industrial applications. It was "grafted" onto an existing railroad network for delivering coal to steam engines, but with increased demand for power, a new AC-based infrastructure had to be built. George Westinghouse saw this change coming and took advantage of the opportunity. He invested his inventive and entrepreneurial energy into the emerging power architecture and set the stage for a new phase of industrial and residential growth in America.*

Soon, the General Electric Company of Schenectady, New York, became the biggest beneficiary of this system development, because it had sufficient financial capital and technological prowess to harness the growth. Following the example of George Westinghouse, the company bought patents and hired the best inventors, such as Charles P. Steinmetz and William Stanley, who helped GE create strong technological advantage in the emerging electric power industry. Working on GE's assignment, Swedish engineer Ernst Danielsen and American electrical engineer-scientist Luis Bell successfully designed an electric AC motor that circumvented Tesla's patents owned by George Westinghouse. In the early 1900s, GE set up an R&D lab "for commercial applications of new principles, and even discovery of those principles" [22].

* Samuel Insull (1859–1939) was instrumental in creating the government-regulated electric utility business model, which evenually became dominant worldwide.

Distribution Buildup is a period in a system's development curve when we have to address constraints and trade-offs accepted earlier, due to the initial lack of resources and knowledge about the emerging market. During the Synthesis and Early Growth phases, the system has to prove its viability within the original infrastructure, or perhaps slightly outside of its boundaries. As the new technology moves from the early adopters to the early majority of users, system growth runs into infrastructure limitations.

For example, in Edison's case, consumers had to put up with waiting lists for a light bulb installation and the necessity to live in proximity to a coal-fired power station. Today, e-books can be delivered quickly over a 3G network, but as the books incorporate more graphics, video, 3D models, zoomable user interfaces (ZUIs), and real-time user interaction options, the network will become a constraint.

> While radio waves are all around us, they're not abundant enough for mobile-phone companies. The U.S. government treats airwaves as a public good, dividing them up and selling licenses to companies, including wireless carriers, to use for their communications. These licenses are becoming tougher to find, and more expensive, as the Federal Communications Commission runs low on new frequencies it can offer for sale. [23]

In another example, when Apple first introduced the iPhone, the company positioned the device as a cool phone with a new intuitive user interface, a web browser, instant email access, and a limited set of applications. In 2007, running a YouTube video on a first-generation iPhone over ATT's Edge network was not a great user experience. The device had to succeed despite the constraints of the existing mobile phone networks, which were poorly suited for high-speed data transfers. By 2011, after several years of rapid growth, telecom companies realized that a new era in mobile applications had arrived. Their massive investment in 4G (LTE) networks alleviated the earlier bandwidth constraints. (This improvement in bandwidth can give rise to a different constraint in a component of the system, such as battery life of the mobile device working on a 4G network, storage capacity, etc.) By 2013, as the demand for mobile data services continued to grow, wireless spectrum turned into one of the industry's most precious resources.

REVIEW QUESTIONS

1. Which of the recent Google business acquisitions can help the company succeed during the Distribution Buildup phase?

   a. Motorola Mobility, a mobile device manufacturer

   b. Frommer's, a publisher of travel guides

   c. BufferBox Inc., a service for parcel pickup stations for packages ordered online

   d. RightsFlow, a provider of music licensing services and royalty payment solutions

   Explain your choice. Sketch out system diagrams for each of the acquisitions. What system elements does each company implement within its system?

15 Growing Up
A Paradigm Shift within the System

The transition from an old to a new infrastructure represents an important shift within a technology and business environment. Before the shift, the original system has to succeed *despite* its environment. That is, the early adopters have to put up with relatively unreliable devices, limited services, high adoption costs, and other inconveniences. After the shift, the environment is being built to make the system as successful as possible. That is, when the new opportunity becomes obvious to major industry players, they move in to encourage and capitalize upon consumer demand.

For example, creation of a reliable, post-Westinghouse electricity distribution system spurred enormous growth in the use of electric devices. Architects designed new buildings with access to electric power built in. Improved incandescent light bulbs, irons, toasters, and refrigerators became ubiquitous in homes, offices, and factories. Entrepreneur Henry Ford found a way to apply electric motors for powering his mass manufacturing system. Electric metalworking machines, such as a lathe with a built-in motor, began replacing steam engine–driven line shafting equipment. Among other novelties, new entertainment industries emerged, such as radio broadcast networks and movie theaters with air conditioning. Neither one of those could exist without a widely available electricity distribution infrastructure. After people started believing in an electric future, such future became inevitable.

In another example, the automobiles of the early twentieth century started succeeding despite many obstacles, such as high maintenance costs, frequent accidents, the lack of road infrastructure, and speed limits of 12 mph imposed by cities concerned that the new vehicle would scare horses and kill pedestrians. But once the automobile proved its usefulness, public and private parties began making investments to facilitate its widespread use. As an illustration, the largest public works project in United States history was the creation of a national highway system authorized by the Federal-Aid Highway Act of 1956. The law provided billions of dollars of public money to build 41,000 miles of high-speed motorways for automobiles. The governments of Germany, China, Japan, India, and other countries invested in road building on a similar scale.

Furthermore, private businesses built gas stations, residential areas, shopping malls, office parks, and many other commercial objects to enable convenient use of the automobile. Most of today's infrastructure in large parts of North America, Europe, and Asia was created with a built-in *expectation* of car-based transportation. The automobile is a successful innovation because over a period of time the society invested a lot of resources to make it so.

Today, automated and driverless cars are being introduced into an environment that makes it difficult to deploy them on a large scale [24]. Nevertheless, we can fully expect that once the robotic cars start gaining broader consumer acceptance, roads, and other infrastructure elements will be changed to give them additional advantages over current driving methods.

Similarly, the original shopping cart (Figures 15.1 and 15.2), a 1937 invention by Sylvan Goldman, succeeded despite the inconvenience of using it in

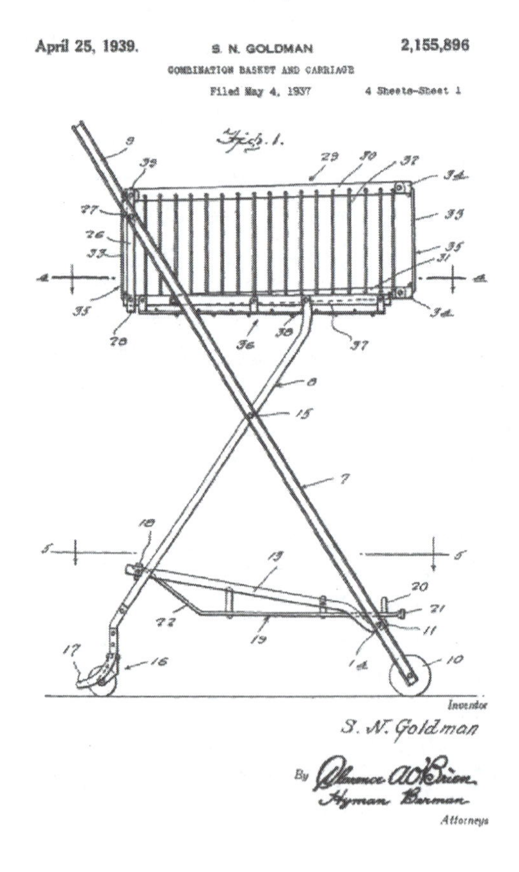

FIGURE 15.1 Shopping cart, US Patent 2,155,896 [25].

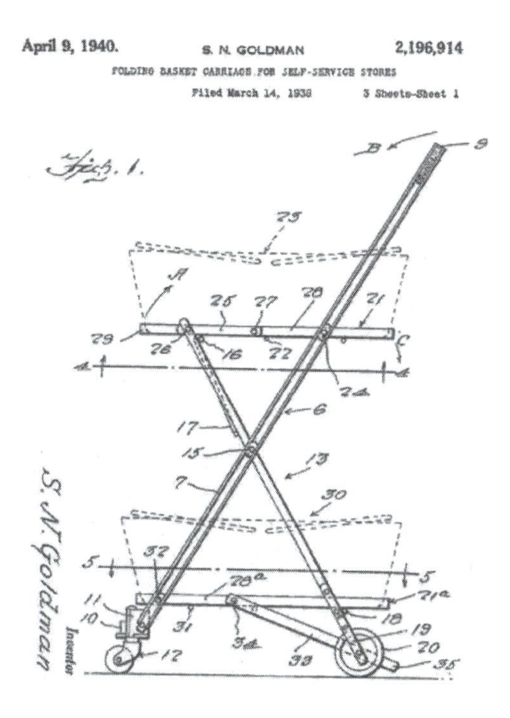

April 9, 1940.
S. N. GOLDMAN
2,196,914
FOLDING BASKET CARRIAGE FOR SELF-SERVICE STORES
Filed March 14, 1938
3 Sheets-Sheet 1
Fig. 1.
S. N. Goldman
Inventor

FIGURE 15.2 Shopping cart with multiple baskets, US Patent 2,196,914 [26].

a small shop, and in narrow spaces between aisles of produce and merchandise. Because at the time of the original invention most people did not have cars, buying a lot of goods with a shopping cart meant additional inconvenience in taking one's purchases home. Nevertheless, over time all these obstacles were removed; new supermarket infrastructure and goods delivery systems were created to facilitate cart-based shopping. As a result, the shopping cart, both as a physical carrier of purchased goods in a store and as a virtual collection of items to be ordered online, became ubiquitous.

Note that Sylvan Goldman designed his cart so that shoppers could fill it with a lot of physical goods. As an extension of this idea, users can fill the virtual shopping cart with a lot of *aboutness* items, which point to products, services, and experiences. Among other things, the aboutness can be used for instructing warehouse robots to assemble the orders and send them to their destinations. Amazon's acquisition of Kiva Systems indicates that the cart's "luck" will last far into the future.

It is easy to predict a similar shift in the e-book business. At the time of this writing, most of the infrastructure—libraries, bookstores, schools, universities, and other institutions of knowledge—is built with the assumption that the paper book is the dominant design for the Packaged Payload. Even electronic repositories (a form of libraries for scientific publications, which now extend well beyond literature in its strict definition) provide the vast majority of articles in formats better suitable for either printing or reading on a personal computer. Nevertheless, we can see key industry players getting into the game. Amazon is creating a self-publishing infrastructure and enabling public libraries to lend out e-books. Apple is building an e-publishing platform tied to its second-generation iBook application and cloud-based iTunes services. We haven't gotten out of the Early Growth phase yet, but the trends point toward the inevitable transition toward an e-book-oriented knowledge packaging and transfer infrastructure.

The process of infrastructure buildup is not problem-free, though. It inevitably requires massive capital investment and often results in overspending. That is, the Distribution has to be built to accommodate new instances of Sources, Packaged Payloads, and Tools, which are yet to come in the future. Often, this growth does not follow within a reasonable (expected, required, desired—for reasons of economic payback, for example) period of time. In this case, the increase in the Distribution capacity enabled by infrastructure investment produces no return. In the most recent example, during the aftermath of the dot-com bubble, major communications companies Worldcom and GlobalSecond went bankrupt. AT&T, one of the oldest telecommunication companies in the United States, had to reorganize and was eventually acquired by SBC Communications. Even when infrastructure buildup is performed through government investments or public–private partnerships, there is a high probability of overspending and financial failure.

16 Infrastructure Innovations *Timing Is Everything*

Because solving infrastructure problems requires large investments, people tend to consider all large-scale investments as a Distribution Buildup phase that necessarily enables future growth. Let's use the system model and the S curve to compare and contrast two recent technology developments labeled as infrastructure projects.

The first one, Project Amp, involved the US government's efforts to jump-start the widespread use of solar energy. On June 23, 2011, cnet.com reported [27]:

> The Department of Energy has issued a $1.4 billion conditional loan guarantee to fund a massive project that would install solar panels on unused industrial roof space across the U.S. Different from many private solar roof installations, the electricity generated from the Project Amp solar panels will feed directly into a branch of the national electric grid, not the host building itself.

The second, Google Fiber, is designed to discover new high-bandwidth Internet applications. On September 27, 2011, cnet.com reported [28],

> Google has begun building its free high-speed network, which promises speeds of 1 gigabit per second for both downloads and uploads using fiber-optic lines to the home. "We believe the uplink capacity is the real game-changer here," Lo said. "We're going to light up our customers in the first half of next year." Even though the company is confident the project makes business sense, it's not clear what exactly gigabit Internet will bring to households beyond some ideas such as videoconferencing.

How do these two large-scale investments compare in terms of creating opportunities for growth? To answer this question, let's put together a system model for each case and determine where on the S curve the system belongs.

TIMING EXAMPLE 1: $1.4 BILLION FOR ROOFTOP SOLAR POWER PLANTS

First, what functional role does a solar power plant play within the system? Is it an instance of Tool, Source, Packaged Payload, Distribution, or Control?

As we discussed earlier, for example, when we considered a system model of Edison's electricity distribution system, the power plant functions as a Source. It generates an instance of the grid-compatible Packaged Payload in the form of high-voltage AC power, because the electric grid in the US is AC based. The new Source is added via Distribution (the existing electric grid) to feed power to homes, offices, commercial facilities, factories, and other Tools. The Control system, that is, the way electricity is directed toward Tools via specific Distribution routes, is not significantly affected by the change. The addition of new Sources means that more electric power can be used within the system. Alternatively, older Sources can be replaced with the new ones without increasing the overall power use or pollution.

With the basic system model in mind, let us determine where the system is on its S curve. For example, are we in the Synthesis stage? The answer seems to be *no*, because all other elements of the system remain the same. Just in case, let's run a quick check:

Tool: No change in the way we use the electric power.

Distribution: No change in the way we transmit power.

Packaged Payload: No change in the kind of power we use—same high-voltage AC on the wire.

Control: No significant change in system setup or performance orchestration; a minor investment is necessary to accommodate the mismatch between power demand and the solar cycle.

Are we in the Early Growth stage? No, because there are no Tools added to the system to take advantage of the new quantity of the electrical power that is becoming available. Also, we are not in the Distribution Buildup stage because the grid as a whole is not affected. And, since there's no new grid expansion, totally new uses of power are not available either.

To summarize, this particular infrastructure investment doesn't seem to enable future growth because it does not affect the system's bottlenecks and does not create new ways to use electricity.

Evaluating the Green Energy Argument

Let's consider the argument that Project Amp provides for future growth by eliminating greenhouse gas pollution. That is, by replacing conventional power plants running on fossil fuels, the investment removes an external constraint imposed on growth by global warming.

The information to support this claim is an estimate of how much growth the new power can support. For example, Project Amp when completed is said to be able to power 88,000 homes [29].

In other words, the growth will come when we find additional investment to construct those buildings. Of course, there was no such growth pending the removal of the power constraint. Moreover, after the burst of the housing bubble, with residential foreclosures and office vacancies at record levels, the demand for those thousands of houses is shifted several years into the future. Since the solar power technology is still improving, as a result of this infrastructure investment we get the worst of both worlds: a long-term commitment to a relatively costly, outdated technology and no growth in sight. The current grid is a very mature system built for the industrial age; therefore investment in additional power sources that are hard to control with old means is likely to create more strain instead of growth.

A much greener and more economical approach would be to focus on a new energy S curve and to try to solve a Synthesis problem. One way to do it would be to localize solar power generation, storage, and usage—a more distributed model. Furthermore, investment in improved high-capacity batteries could allow for more local storage, reduced grid load, and even entirely off-grid installations. These solutions are likely to be in demand not only in the United States and European Union, where solar energy is heavily subsidized by the taxpayers, but also in developing countries, whether in cities where massive pollution accompanies rapid growth, or locations where the electric grid is unreliable or nonexistent.*

For example, Boston Power, a US startup producing longer-lasting high-capacity electric batteries found strong support for its technology among Chinese automotive companies addressing local market conditions [30]. To reduce air pollution, the mayor of Beijing, the capital of China, allows electric battery-powered cars to be on the streets seven days a week. On the other hand, to drive a gas-powered automobile in Beijing, one has to enter a lottery that lets the winner drive four days out of seven. The innovative system creates strong incentives to scale up the use of "green" cars. In India, where in 2011 a quarter of the population had no reliable access to electricity, solar energy is now successfully competing with diesel generators, despite the fact that solar panels cost more than the generators [31]. With diesel-generated energy production in India estimated at 30 gigawatts, there is a large potential for economically effective green energy Synthesis solutions [32].

* Alternatively, an investment into a smart grid, while copiously funding basic research into solar panel efficiency, could produce synergy when integrating the system components.

These and other examples show that investments in green technology, when directed at a problem that matches the system's position on the S curve, can produce sustainable growth while improving environmental conditions. In contrast, due to a system-level mismatch, the money spent on Project Amp, where new green Sources are added to the old Distribution, would not lead to comparable growth.

TIMING EXAMPLE 2: THE GOOGLE FIBER PROJECT

From the system perspective, Google's new high-bandwidth network plays the role of Distribution, because it enables delivery of large amounts of information from data centers and other users to homes, offices, and commercial locations, where it can be used in new bandwidth-hungry applications. The project aims to discover such applications—new physical and virtual Tools—and connect them to new or existing Sources.

Where on the S curve does the project belong? Since its purpose is to dramatically increase bandwidth available to additional user devices, it can be either in the Synthesis or the Distribution Buildup phase. For Synthesis, we would have to add a completely new type of a networked device (Tool), a new type of content (Packaged Payload), new or dramatically improved Sources, and new ways of orchestrating their performances (Control). So far, any evidence of such radically new developments is lacking. At the time of writing of this book, the Google Fiber blog describes the effort as "installing fiber and building the Google Fiber Huts" [33]. According to the diagram provided by Google (Figure 16.1), Fiber Huts serve as interfaces between the global Internet connectivity network and local fiber networks.

Although the emergence of a high-performance interface is important from a system perspective, it doesn't signify Synthesis of a new system yet. (Unless Google has the requisite components ready to be deployed once the Distribution component is in place. Should Google and its partners succeed

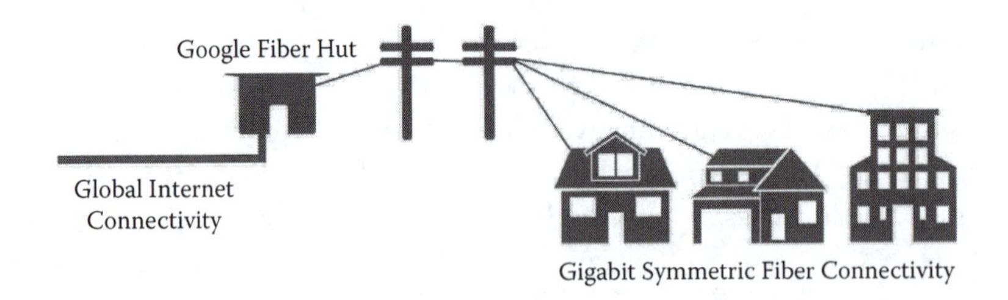

FIGURE 16.1 Google Fiber project diagram. (From a construction update, *Google Fiber Blog*, April 04, 2012 [33].)

with the deployment, the interface will become a Control Point within the emerging digital content distribution system.)

On the other hand, the installation of multiple new optical fiber channels for sending and receiving massive amounts of data indicates a Distribution Buildup. This development would be particularly beneficial to digital video because Google Fiber removes bandwidth constraints that put today's Internet TV at a disadvantage relative to digital cable TV. That is, in the new system, which reportedly is going to have data transfer rates up to a Gb/sec, any high-definition movie selected for viewing would arrive instantaneously. At this level of performance, the distinction between streaming and downloading disappears. In fact, this level of performance would be similar to today's TV channel change experience—no waiting for content download is necessary. In such an environment, high-resolution digital TVs and set-top boxes are going to become very lucky.

Google Fiber has another feature that can add possibilities for new applications and devices to the existing Internet user scenarios. Unlike most of the current asymmetrical broadband systems where download speeds are about ten times faster than upload speeds, the system being created by Google will be symmetrical, with equal upload and download speeds. This means that high-bandwidth user content will have a much easier time getting to data centers and other users. For example, video conferencing, graphics-rich gaming, telepresence, lifelike online collaboration, and other two-way networking apps could become an everyday reality. Whatever the final result of this network buildup, it will lead to increased use of networked devices on both sides of Google Fiber, thus enabling systemwide growth.

In sharp contrast, the $1.4B Project Amp directed toward solar panel installation in remote places can produce incremental changes at best. Positioning Project Amp as an infrastructure project creates a false sense of potential growth, which it is not going to deliver beyond the immediate infusion of money into the economy (i.e., the hiring of a certain number of electricians and the purchase of a finite amount of materials and supplies). The investment is unlikely to generate additional business or technology opportunities. It is a solution with a built-in mismatch with the system as a whole.

17 Infrastructure and Growth

Zooming In on the Micro Level

Our earlier discussion about the future of mobile devices in general and e-books in particular had a number of hidden assumptions. That is, we assumed that digital displays, memory, network bandwidth, and processors would become exponentially faster, cheaper, and more abundant. These expectations have proved to be true for at least the past forty years. This amazing track record of the information technology (IT) industry to deliver higher and higher levels of performance at lower prices can be illustrated by an old joke attributed to Bill Gates, "If GM had kept up with technology like the computer industry has, we would all be driving $25 cars that got 1000 miles per gallon" [34].

What underlies both the joke and the real IT experience is the so-called Moore's Law, which in its most popular form states "the number of transistors on a chip will double approximately every two years" [35]. The law originates from a 1965 article in *Electronics* magazine by Gordon Moore, then the director of Research and Development Laboratories at Fairchild Semiconductor. While observing technology trends in the nascent semiconductor industry, he noted that "by 1975, the number of components per integrated circuit for minimum cost will be 65,000" (from 50 in 1965. Figure 17.1). Moore reasoned that no fundamental physical barriers existed to limit the growth, at least in the short term, and "Only the engineering effort is needed" [36] to succeed.

As we can see, Moore's Law is as much about the ingenuity of engineering effort as it is about laws of nature. Since 1965, Moore's insight has become a guiding principle and a pacing parameter for engineers and scientists who develop new materials and computing systems. Like any other self-fulfilling prophecy, the law works because people believe they can make it work. Though claims about the impending doom of Moore's Law circulate on a regular basis, there are strong indications that the trend will continue for at least the next ten years. For example, in 2011 Intel began chip production using 22-nm processes, with 10 nm reachable by 2015 (*nm* is short for

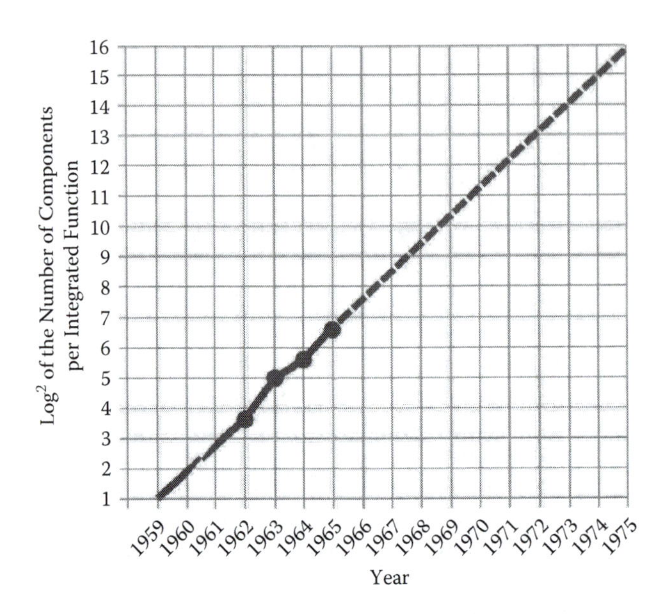

FIGURE 17.1 Moore's original projection from 1965 to 1975. (From Gordon Moore, Cramming more components onto integrated circuits, *Electronics* 38, 1965.)

nanometer, or 0.000000001 of a meter, or 0.00000000328 of a foot. For comparison, if a 6-foot man were 20 nm tall, a dollar bill would appear to him as thick as Mount Everest, the highest mountain on Earth.) Moreover, in January 2012, IBM researchers announced that they managed to magnetically store and retrieve a bit of information using an array of just 12 atoms, about 5 nm wide. The continuing technological successes of the last 50+ years make some futurists believe that by 2045 computers will become more intelligent than people and begin developing themselves at an even faster rate than humans ever could [37].

Psychologically, the shrinking size of individual components, such as the 20-nm transistor or atom-based storage, attracts the most attention. That is, studies of human emotion show that we feel awe when dealing with enormously large or infinitesimally small objects [38]. Nevertheless, from a system perspective, we can see that in connecting lots and lots of transistors in a meaningful way, synchronizing their work would be as important as making them incredibly small. In Moore's early days, 50 components on a chip was a reasonable number; projecting that number to 65,000 in ten years sounded like science fiction. But the majority of modern computing applications involve chips containing millions, and even billions, of individual components. This exponential growth in the number of connected elements hints at a major Distribution invention, hidden in the popular definition of

Moore's Law, but mentioned explicitly by Moore himself in his original 1965 article.

The invention is called the integrated circuit (IC). It was created by Jack Kilby of Texas Instruments [39], and independently, six months later, using a different implementation, by Robert Noyce, of Fairchild Semiconductor [40]. Kilby's work won him 50 percent of the 2000 Nobel Prize in Physics, the prize he split with Zhores I. Alferov and Herbert Kroemer, the physicists who, according to the Nobel Prize Committee, "have invented and developed fast opto- and microelectronic components based on layered semiconductor structures, termed semiconductor heterostructures" [41].

Both Kilby and Noyce invented a way to integrate several electronic components, including transistors, within the same physical structure. The Noyce solution, implemented in silicon, was more practical and eventually became the prototype for most industrially produced ICs.* In 1968, Robert Noyce and Gordon Moore left Fairchild Semiconductor to found Intel Corporation, a small startup focused on designing and manufacturing semiconductor memory. In 1971, Intel developed the first commercial microprocessor. But it was not until 1983, when under competitive pressure from the Japanese manufacturers in the memory business, Intel decided to shift into microprocessors, taking advantage of the sudden success of IBM's personal computer (PC). Lucky microprocessors!

By the end of the 1980s, as the number of transistors on an IC and their speed of operation continued to grow exponentially, the interconnect between elements on the IC, that is, Distribution in system terms, started turning into a bottleneck. To appreciate the problem, let us consider the earlier nanoscale example with a 6-foot man and Everest. Using the same proportions, an inch-long microchip would be equivalent to the distance between San Francisco and Oklahoma City, a three-hour flight halfway across the United States. Because transistors have to work in unison, imagine the man in California having a dance session with his girlfriend in Oklahoma. The faster and more complex the dance, the more difficult it is to synchronize the pair's moves. Furthermore, using the scale-time-money (STM) operator, imagine a million couples like the first one dancing at the same time in the same place. Then, imagine a billion people, almost the total adult population of China, dancing simultaneously. It is clear that with the growing speed and complexity of the dance, as well as the density of people on the dance floor, communications between individuals and groups of dancers become essential to the success of the whole performance.

* Kilby was a great inventor, but Noyce and his team turned out to be better innovators.

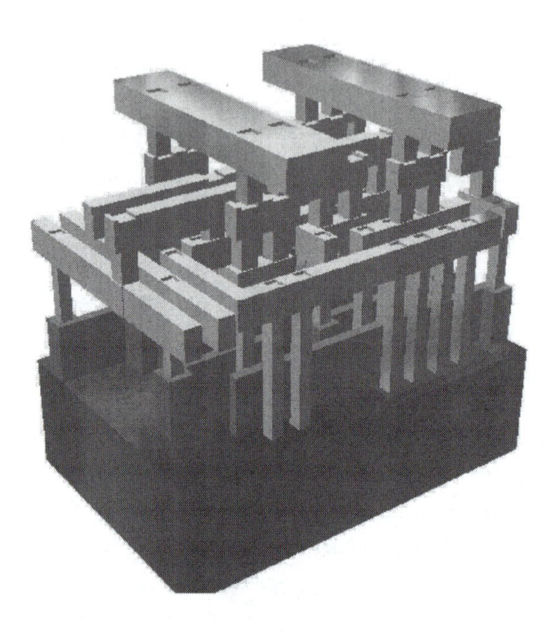

FIGURE 17.2 3D representation of a multilayer IC metallization scheme. (From http://en.wikipedia.org/wiki/Integrated_circuit.)

Organizing and connecting such elements with wires represents a difficult challenge to engineers.

During the 1990s, the challenge of the interconnect was addressed by a number of technological advances in IC design and manufacturing technologies [42]. Among the breakthroughs, engineers and scientists found ways to develop multilayer metallization schemes (see, for example, Figure 17.2), which allowed for the segmentation of functional IC parts. This solution helped simplify interaction between elements within the same IC segment, interaction between different IC segments, and modular design of a particular IC depending on computing requirements.

Overall, the semiconductor industry advances of that decade enabled a wide user adoption of computation-intensive PCs and workstations with graphical user interfaces (GUIs) and the development of high-performance servers, built using general-purpose microprocessors. Both the PCs and the servers were important system building blocks during the years of the Internet revolution of the late 1990s and the early 2000s. One innovation process fed another. The demand for GUI applications and web pages created demand for continuing growth in IC performance according to Moore's Law.

At the time of writing of this book, the IC interconnect is facing its next challenge. Most of the advanced chips today, either in PCs, servers, or tablets, contain multiple computing cores running at frequencies between 1 GHz and

4 GHz. To translate the numbers into device performance, each chip core can execute from one to four billion instructions per second. The greater the number of cores, the more transistors they contain, the faster they run, and the more data has to be moved in and out of the chip and between cores. Because in electronic devices data is represented by electric currents, quick information transfer requires equally fast electron transfer over a large number of conducting paths. The transfer causes resistive and other types of losses. As a result, the greater the number of cores and the faster the communications speed we want to achieve, the higher the losses and synchronization delays. According to some estimates, interconnects consume 80 percent of a microprocessor's power [43]. Just like before, the Distribution component is becoming a bottleneck for growth in IC performance.

Many engineers and scientists believe that they can solve the problem by using optics instead of electronics [44]. That is, to convey signals, one could move photons instead of electrons. The approach would allow integrated circuits to reach data transfer speeds above one trillion bits per second (Tb/sec) with minimal energy losses. From the mid-2000s, "optical links are already being used to connect racks of equipment that span 1–100 m" (Figure 17.3 [45]). But the scaling of an optical interconnect down to practical on-chip applications still seems a few years away [46]. Nevertheless, even the critics of the technology acknowledge that "None of these are challenges are insurmountable."

Whether the new micro-level interconnect breakthrough will be achieved based on optics or on some other awe-inspiring discoveries in physics, such

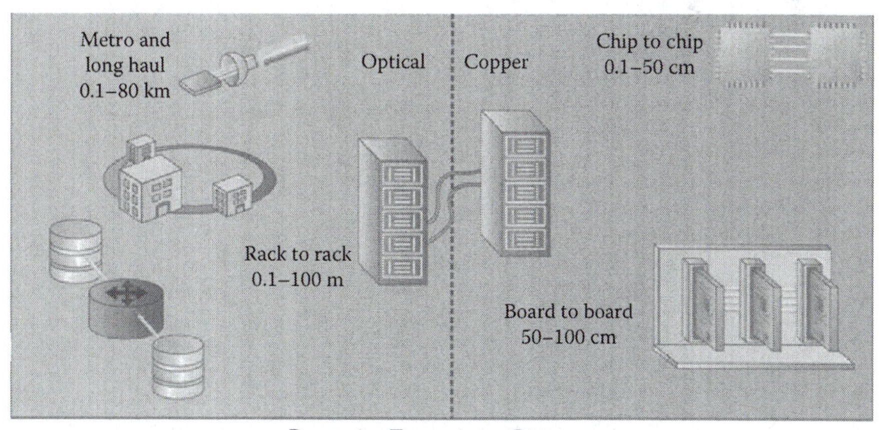

FIGURE 17.3 Optical versus electrical interconnect. (From Andrew Alduino and Mario Paniccia, Interconnects: Wiring electronics with light, *Nature Photonics* 1, 2007 [45].)

as graphene or carbon nanotube transistors [47], we can be sure that a solution to the high-value Distribution problem will spur new growth. It will happen through the creation of new computing systems, rapid expansion of the existing ones, or more likely, both. In the meantime, understanding the relationship between the S curve and evolution of system elements will allow us to monitor the developments and anticipate potential growth, all without succumbing to hype and false expectations.

18 A Stage of System Evolution
Efficiency

The Distribution Buildup phase creates multiple opportunities for growth. As the new phase ramps up, more instances of Sources, Tools, Distribution routes, Packaged Payloads, and Control elements are added to the system. Further, the instances become specialized, targeting a diverse range of specific applications. This type of growth requires better coordination; therefore, the Control functionality begins to play an increasingly important role in providing for the overall system performance. Information exchanges between system elements, extraction, and analysis of data, generation and distribution of control signals, and accumulation of relevant information for further processing turn into stable, routine processes. The system enters the *Efficiency* stage where growth is supported by setup, anticipation, and real-time performance orchestration, rather than simple increase in the resources available to the system.

As inventors and innovators, we can use a simple rule of thumb: The more instances of various elements that are present in the system, the more important the Control is. For example, as we discussed in Chapter 3 the more songs, books, movies, podcasts, and other content Bob has in his media library, and the more media sources and play-out devices he can access, the more valuable his iTunes setup becomes to him. In a different example, the more items a retailer, such as Walmart or Amazon, wants to put on the shelves in their virtual or brick-and-mortar stores, the more important their merchandizing, marketing, and supply chain management systems become. In other words, in the Efficiency stage, the system devotes more resources to ensuring that the right Packaged Payload gets from the right Source to the right Tool over the right Distribution route at the right time and at the right cost.

Consider Google's success with its self-proclaimed mission to organize the world's information—a system-level Control function. The more web pages and documents that are out there reachable by Google's indexing tools, the more valuable Google's ability to deliver the right piece of information—and advertisement—to the right consumer becomes. While Yahoo failed to scale with its catalog-oriented approach, Google succeeded with relevancy-based search. As the web and network bandwidth grew—from dial-up to broadband to optical fiber—so did Google's reach into other types of web-related media: video, maps, images, books, news, and so on. What was good for the free flow of web content was also good for Google's advertisement delivery and web analytics business.

Nevertheless, the emergence of Facebook as a semiclosed social network, the shift from the web to mobile applications, Apple's iTunes and App Store, as well as other postweb technology developments makes it more difficult for Google to collect user information and deliver the right ads. The web giant can't organize what it doesn't see. To stay relevant, Google has no choice but to battle with Apple, Facebook, Amazon, and others to provide Control functionality in the content creation and distribution system. Because Apple and Facebook (especially Facebook) have access to a lot more personal user information, they are becoming strong contenders for advertisement dollars. Today, Google is focusing on social networking (Google+) and mobile devices (Android) because these new technologies can help the company access more information, learn more about consumers, and deliver fine-tuned commercial payloads.*

SYSTEMWIDE ELECTRICITY DISASTERS

Of course, the need for the Control functionality exists during the Early Growth stage, but since the variety and the number (or volume for the Packaged Payload) of element instances is relatively small, simple control mechanisms can cope with efficiency-related problems. For example, in Edison's systems there were multiple power meters and tracking mechanisms, but usage scenarios included either lighting or simple electric motor applications. In contrast, when the electric grid matured in the post-Westinghouse era, it was powered by a wide variety of electric stations that were constructed over a long period of time and used different power sources, such as hydro-electric, coal-fired, gas-fired, and nuclear power. The stations had different output capacities, ranging from 1,000,000,000 (mega-) watt to 1,000,000,000,000 (giga-) watt. Accordingly, the stations were connected to the grid with power lines of different current-carrying capacity. To deliver the right power to the right customer—industrial, commercial, or residential—the right substations with transformers had to be set up. Since energy usage varies significantly depending on the time of day and weather patterns, controlling the flow of electricity over the grid is an essential task for electric utility companies. Failure to do so can result in hardship for millions of people and inflict huge financial losses.

---

* For an in-depth exploration of how competition among Google, Facebook, and other industry players shapes the evolution of the system, see Chapters 20 through 22.

For example, according to the *New York Times*, the blackouts in New York City of 1965 and 1977 turned into defining moments in the city's history. Both times, the blackouts were caused by a failure of a regional utility to manage a sudden variation in electricity flow. In 1965, a bad breaker setup on one power cable resulted in a power surge that overwhelmed the city [48]:

> In New York City, as monitoring dials went wild, a crucial operator at the main power control station made frantic phone calls upstate instead of preemptively shutting down the city's system to protect it.

Due to the sudden surge, automatic shutoffs occurred, affecting 25 million people during rush hour; people were stuck in elevators, roads, and even airplanes.

In 1977, a series of three lightning strikes triggered a cascade of intentional and unintentional power shutdowns in New York City. As the operators struggled to contain the damage and bring the system back to life, incidents of arson, looting, and vandalism occurred in 31 neighborhoods, resulting in 3,776 police arrests. The total damage from the power outage was estimated at $300 million.

In 2003, the Northeast Blackout occurred, leaving 50 million without power and inflicting an estimated $6 billion in damages. The initial failure of a high-voltage line in the state of Ohio was caused by overgrown trees. Unfortunately, due to a software bug deeply embedded in GE's energy management system [49], the regional control facility didn't handle the problem in time. As the result of this and other cascading failures, people in eight US states and parts of Canada lost their electric power [50].

As we can see, in a large mature system, even relatively minor failures of the Control function coupled with limited Packaged Payload storage capabilities (we are yet to learn how to economically store large amounts of electricity) can trigger huge systemwide disasters.

EFFICIENCY AND THE FUTURE OF THE E-BOOK

With regard to the e-book, we are barely past the Synthesis stage, entering Early Growth. Before it gets to the Efficiency stage, the system would have to get at least through the Distribution Buildup phase, and the emergence of new types of instances of all major system elements: Sources, Tools, Packaged Payloads, Distributions, and Controls.

As we discussed earlier, the Packaged Payload interfaces with all other system elements. To develop long-term transformation scenarios for the system as a whole, we need to consider trends in Package Payload–related technologies. Applying these considerations to the evolution of books, which are going to be produced, distributed, and consumed digitally, we need to think of them in a broader context of major technological developments affecting electronic documents in general.

19 Payload Evolution
From Physical to Virtual

If technologies were people, they could tell us their life stories, so that it would be easier for us to appreciate where and how they grew up and what influenced their development. We can start a life story of the electronic document with a key transition that occurred about fifty years ago when computers obtained displays as output devices, making digital documents visible and interactive. It is hard to imagine now, but at the time people were mesmerized with their new ability to change a document before printing it. Unlike the typewriter, a computer program could let users edit texts on the fly. For example, instead of using scissors, glue, and strips of paper, you could copy and paste words electronically. It was a strange but exciting new use for the digital computer, originally designed as merely an advanced calculator.

Electronic documents at the time comprised flat text, that is, a sequence of designated digital codes corresponding to letters, numbers, and nonprintable characters. For example, number 65 represented the letter "A," 66 the letter "B," and so on. To ensure compatibility between different computer systems, the character-encoding scheme was standardized as the American Standard Code for Information Interchange (ASCII), a standard based on codes used earlier for telegraph communications. The interaction with the document on the screen was accomplished using a keyboard, jumping from letter to letter or line to line by pressing arrow keys, and later, combinations of keys, some of which, like Ctrl+C for copy and Ctrl+V for paste, are still used on modern computers. (Remember that one of the loudest complaints against the first iPhone was that users couldn't copy and paste easily.) Despite the great convenience for editing, when printed, the document looked much worse than a published or even typewritten page. Nevertheless, the new ugly computer-based system had a huge potential advantage over the old beautiful paper-based one: People could update the text quickly and print several modified copies on demand.

The next big change came in the mid-1980s, with the commercial success of the Apple Macintosh personal computer, which featured a graphical user interface (GUI) and a computer mouse. The Mac was a direct descendant from computing principles and machines originally developed by Douglas Engelbart at the Xerox Palo Alto Research Center (PARC) in the 1960s.

Other important developments at Xerox PARC were the laser printer and InterPress, a programming language for generating printer commands from a digital document file. At the time, fonts were not widely available on all computers. Moreover, the printer capabilities and commands they could execute varied widely. Therefore, a document created on a computer with high-quality fonts and other advanced text formatting features could not be transferred to a less-capable computer system without a major loss of quality. To solve the problem, the new programming language allowed users to print to what we would call today a *virtual printer*, creating a specially encoded file. Then, the file could be transferred to any computer and printed to any compatible printer. In 1982, researchers from Xerox PARC, who couldn't get the InterPress technology successfully commercialized inside a large corporation, founded startup called Adobe Systems. The startup created the PostScript language, and with Steve Jobs' encouragement, ported it to the Mac and Mac-compatible laser printers. Some believe that desktop publishing saved the Mac from being a commercial disaster [51].

One of the important GUI principles perfected by Apple was *what you see is what you get* (WYSIWYG), a method that allowed users to see the formatted document on the screen in the same way it would come out later in print. One of the few courses Steve Jobs, the cofounder of Apple, took seriously in college was calligraphy. Learning the aesthetics of exquisitely written letters helped him later to craft and select fonts for his new GUI-based computers. The new system gave users a variety of new text formatting options, including font change, collation, and so on. While most computer systems at the time offered users a trade-off between the ease of editing an electronic document and the document's ugly look when it was printed, a Mac connected to a laser printer gave them the best of both worlds: ease of use and professional-looking pages. This breakthrough, in combination with intuitive cursor navigation enabled by the mouse, allowed nonprogrammers to use personal computers for most of the desktop publishing tasks we take for granted today: from editing to layout to picture insertion.

Furthermore, with the proliferation of computer spreadsheets, documents became *live*. That is, instead of three distinct steps of entering the data into the computer, then running number-crunching software, and later reviewing the results on screen or in print, the user acquired the ability to manipulate numbers and formulas in real time. The visual aspect of this innovation figured in the name of the first commercially successful PC spreadsheet application VisiCalc. Bricklin and Frankston wrote the software for Apple II and later ported it to the IBM PC. As the result of these and other technology breakthroughs, including color printing, personal computers became document-producing factories, both in electronic and paper forms of the medium. The desktop publishing world was born.

The large scale of document production created the problem of how to organize and share them—a typical situation in system evolution, when a dramatic increase in productivity of Sources and Tools puts pressure on the Control element. With hundreds and thousands of files on one computer, and millions of them within a university or a corporation, management of information turned into an important, but often nightmarish task. The typical way to deal with the problem was to create virtual file folders similar to the paper folders used by experienced bureaucrats for centuries before. The method worked (and still works) reasonably well for one person on one computer, but it started failing as computers became connected by a network (Distribution). That is, once the network allowed users to share and modify documents, different computers could now host different versions of the same document, with some users working with obsolete information. Moreover, maintaining coherence of a collection of related documents became a difficult task, because an update of one document may require a manual update of many other documents. We may not even *know* which documents need to be updated and when. The more documents we have, the more connected they are, the faster they need to be updated, the more difficult it becomes to keep the system up to date and under control.*

One solution to the problems of the traditional filing system in a networked world was to organize documents as a logical collection and modify them so that they contained embedded links to other relevant documents in the collection. Hypertext, one such technology for organizing information that dates back to the 1960s, eventually evolved into what we know today as the World Wide Web.

NOTABLE EARLY ATTEMPTS TO INNOVATE WITH LARGE COLLECTIONS OF DOCUMENTS

Tim Berners-Lee, a programmer at CERN, is considered to be the inventor of the web. In 1980 he created ENQUIRE, an early implementation of the hypertext technology, to improve information sharing between hundreds of scientists across multiple departments with disparate computers running different software. The attempt had limited success, mostly due to a cumbersome change propagation mechanism [52, p. 10].

* Note how we apply the Scale-Time-Money (STM) operator to appreciate the scale of an emerging problem. We also use the Three Magicians method to navigate between levels of individual computers, networks, and networks of networks.

That is, each text was represented by a special card from a database, and after changing the text or a linked card, the user had to change related cards as well. The benefit of this two-way design was that you could easily find all documents that were pointing to another document. All you needed to do was examine the backtracking links on the latter page. But the more dynamic the document collection was and the more updates users introduced, the more difficult it was to maintain the cards in working order. Today, to accomplish the same goal, Google and other search engines have to scan and index almost the entire web.

In 1987, Apple released HyperCard, another card-based hypertext application created by Bill Atkinson. The software had a sophisticated link navigation system, its own programming language, and media support. It was reasonably successful within the Apple community, but its future versions could not compete with the web-based technology because HyperCard didn't work across different computer networks.

It wouldn't be fair to characterize these early hypertext technologies as failed attempts to create the web. They simply addressed a different problem: how to organize related documents on a local computer or a network. In the case of ENQUIRE, it was a research institute network, and the scale of the Internet of the 1980s was nowhere near the network of networks it became in the early 1990s. For Apple, HyperCard worked for Mac computers and was compatible with AppleTalk, the company's proprietary network protocol, which was considered sufficient for the existing business and consumer markets. To understand the preweb situation better, let's evaluate the early solutions using the "magic" questions we discussed in the chapter about the Three Magicians method.

First, did the solutions work? The answer would be a qualified *yes*, because the software did what it was designed to do. The user could successfully create documents and link them in a reasonable amount of time.

Second, did the solutions scale? If we consider them on the scale of a computer or a local network, then the answer would be closer to a *yes* than a *no*. But if we rise above those levels to the scale of a network of networks, the answer would be *no*.

Third, was there a need for scale? *No*. Until the Internet became widespread in universities and in some government institutions, there was no need for scaling up to a truly worldwide implementation. Corporate and proprietary networks did a relatively good job in helping knowledge workers share and organize information.

As we can see, the early attempts were not failures to create the World Wide Web. Rather, they were local technology solutions for local information-sharing problems. The situation was similar to the one with Edison's system, where all light bulbs and DC motors performed well enough within their own business and technology networks. On the other hand, building many more or larger systems presented a greater challenge. The recognition of this new barrier led to innovations similar in scope to the ones created by Westinghouse and Tesla for AC electricity systems.

That is, Tim Berners-Lee recognized the higher-level web problem in the late 1980s, ten years after his initial attempts to solve the problem. He happened to work at CERN, one of the major hubs of the early Internet, which by that time was quickly becoming a network connecting diverse smaller networks. Due to his earlier experiences with hypertext and availability of the existing communication protocols developed earlier for the Internet and email, he was able to come up with a new system.

THE BIRTH OF THE WORLD WIDE WEB

If a web page could talk, it would tell us that it was born as a simple, flat text document. This was because every computer and network it wanted to interact with understood the ASCII characters. However, to look like a real document, it had to embed special formatting codes expressed in HyperText Markup Language (HTML). There were special codes for bold, italic, underline, and other formatting tags similar to an earlier and more complex technology called Standard Generalized Markup Language (SGML). The web browser, a special application for presenting the web page, would use the tags to format the page for showing it on the screen.

Moreover, to help users organize texts and insert pretty pictures, the page would carry links to additional resources, such as web pages and digital picture files located on other networks (see Figure 5.4). Those links and references were called Universal Resource Locators (URLs). To fetch the resources, the browser needed to use the HyperText Transfer Protocol (HTTP). The technology itself was rather simple. According to Berners-Lee, "the difficult bit was persuading people to join in. And getting them to agree to all use the same sort of HTTP, and URLs, and HTML" [53].

In 1988, Vincent Cerf who, while at Stanford, was instrumental in creating key Internet protocols, convinced the US Federal Networking Council to allow commercial traffic on the Internet. MCI became the first private company to provide private email service on a large scale. Within a couple of years, companies like America On-Line (AOL), Compuserve, and others followed. The newly opened Internet set the stage for the web to become *lucky*.

In the early 1990s, once engineers and researchers agreed to give the simplified web technology a try, the web began to grow. The earlier two-way link requirement was dropped in favor of a separate repository that could be accessed by other computers connected to the network. The repository was called a *server.* Making and linking pages turned into an easy proposition. Just like in any other initial system growth phase, web growth started through the addition of Sources and Tools. How does one add an instance of the Tool? By downloading a web browser to the PC. How does one add an instance of the Source? By adding HTTP server software to a hardware server connected to the Internet. In most cases, the software was free, because enthusiasts of the Open Source movement developed and distributed it at no charge.

To add an instance of the Packaged Payload, a web page, one had to create an HTML document (with text, links, and formatting instructions) and upload it to the server. Once there, the page was available for the whole world to fetch. Since the beginning, most of the content was also free. For example, Project Gutenberg enthusiasts digitized, formatted, and published on the web many classical texts with expired copyrights. Guides on how to write in HTML or set up an HTTP server were also freely available on the web. A new world of content was being created at a very low cost.

The original Internet backbone infrastructure played the role of the Distribution. Universities, government institutions, and large private companies could connect to the Internet directly, while individuals would have to access it through a service provider, using dial-up modems attached to local phone lines. To have a tolerable early web experience, you would have to have a 9.6 Kb/s connection. For comparison, this is at least thousands of times slower than your average Wi-Fi wireless link today. Depending on the reliability of the dial-up connection, a movie that would take only one second to download over Google Fiber, would take a day or two with early dial-up, running up a huge phone bill in the process.

Despite the relatively low data transfer speed, the web enabled a dramatically better implementation of the Control. First, you could seamlessly navigate a practically infinite library of documents created and linked by other people. Second, you could include elements, such as images, created by other people into your own documents. Third, you could create and share your own organization of documents simply by putting together pages of useful links. With the addition of a simple search engine, you could even start a portal company like Yahoo!, which helped others navigate the web.

From a user perspective, interfacing with the web through a browser application was a familiar mouse-over-document experience. Of course, we wouldn't be able to change the contents of the page while we were browsing. But instead of opening and closing files through various program menus, we could just click on links and surf away. Although making an individual page dynamic (for instance, through basic spreadsheet functionality) would require a major programming effort, the author of the document could easily define its layout and link navigation options. The page itself was static relative to a live document in a word processor or a spreadsheet, but as a whole, the linked collection of documents was infinite and dynamic. To summarize, in the early 1990s, the World Wide Web began its existence as originally designed—an infinite shared repository of documents.

At that time, one could create an e-book using web technology, but the book would not be portable. That is, to read it we would have to carry with us a computer connected to the Internet by a phone wire. Paying for the book over the web would also be a problem because, as we discussed in the aboutness chapter (Chapter 5), the web cookie essential for Internet commerce had not been invented yet. Though many people went ahead and did create many free (mostly technical) e-books using early web technologies, the process of reading online, with the text displayed on a low-resolution computer screen, was not a great book experience. In system terms, the implementation of the Tool was not good enough to compete with the existing paper-based system.

MONETIZING THE WEB: THE ORIGINS OF INTERNET COMMERCE AND USER INTERACTIVITY

The decade between 1990 and 2000 was critical in the biography of the web page. Electronic commerce, new media capabilities, and increased demand for interactive content drove the evolution of the Packaged Payload. During that period, billions of web pages were created. Though some of them retained standard static HTML codes, advanced websites began producing dynamic content. They had pages containing executable scripts, for example, JavaScript, and executables in Sun's Java or Microsoft's Active X. The transformation of the web browser was remarkable. From relatively simple software for interpreting ASCII texts with embedded HTML tags, it turned into an application that ran other applications inside itself.

In 1995, Microsoft Corporation, which initially missed the web boat, released its own free browser bundled with the Windows operating system. This event marked the beginning of a war of attrition with Netscape Navigator, then the undisputed web market leader. By 2000, when Microsoft had won the battle of the browsers, web page technology increasingly resembled that of software, rather than the original tagged text documents. Nevertheless, the mode of user interaction with the page remained the familiar mouse-over-document experience that began in the late 1970s and early 1980s. In a browser window, the web looked like another way to interact with familiar documents.

Following this established pattern of PC-based interaction, people printed cooking recipes, driving directions, stock charts, airline tickets, emails, and many other useful web documents. Despite all the talk about the disruptive nature of the web, traditional computing and networking companies—Microsoft, Intel, Dell, HP, Oracle, IBM, Sun, Cisco—benefited immensely from reusing their older technologies in a new application market. Along with technology stocks, prices for shares in companies like International Paper that produced and marketed paper for printers soared as well.

Nevertheless, behind the growth in pages printed, the paths of paper and electronic documents began to diverge further.* If we considered the web page technology as the dominant design for implementing the Packaged Payload, we would notice a major shift in information *packaging* from HTML to eXtensible Markup Language (XML). One of the problems with HTML was that it had text and instructions on how to format it intertwined within the same file. When a web designer wanted to make a minor update to the page or display the information in a different style, she had to edit the entire page. In addition to that, the meaning of various pieces of information on the page, like date, time, name of a person, gender, or place of birth were meaningful only to live humans; search engines, such as Google and Inktomi, could not extract the information reliably. Furthermore, because of its roots in printing, the logical structure of the document (chapter headings, paragraphs, etc.) was only reflected in page formatting. Somebody's mistake or idiosyncratic use of fonts or headers could throw a text analysis program off base, misinterpreting the contents and miscategorizing text elements. The technology switch to XML allowed web developers to tag the text elements in such a way that computers could recognize the meanings (or semantics) of those elements. Also, document-formatting information was becoming increasingly divorced from static printing, and instead directed toward dynamic user interaction with the screen.

* From a purely business perspective, web-based printing looked the same as PC-based printing. Large technology companies, including HP, were making good money on the new technology trend. They perceived web growth as a burst of innovation that was extending the desktop publishing S curve—a fatal mistake. In contrast, from a system perspective, the printer had to be considered as a Tool in two completely different document distribution systems. The PC-based one was at the end of its S curve, while the Internet-based one was entering a period of growth. Because PC-attached printers were used in both systems, printing-related business temporarily experienced rapid growth. But once web documents became dynamic and mobile networking proliferated, the printing business collapsed. As a result, HP, Kodak, and others suffered heavy financial losses.

A REVOLUTION IN THE MAKING: DETECTING MAJOR SYSTEM-LEVEL SHIFTS IN TECHNOLOGY AND BUSINESS

By the end of the twentieth century, the web page, now twenty years old, could be implemented in many different ways. It could contain good-old HTML code, a script, XML code, Java applets, Active X components, Macromedia Flash scripts, instructions to invoke embedded media players with additional proprietary scripts, and much more. To secure and facilitate commercial transactions, the page could be encrypted, or have an associated *cookie*. As we discussed earlier, a cookie is a tiny text file stored by the browser on your computer that tells the host website what you did and when you did it on the site previously, and how long you spent on a particular product page. From a system perspective, not only did the number of instances of the Packaged Payload increase dramatically, but also the types and complexity of instances the system should be able to handle grew significantly as well.

The system as a whole adapted marvelously to the increase in volume and complexity. On the Tool side, more powerful PCs and more sophisticated browsers with plug-ins were created. On the Source side, a revolution in server architecture occurred. As we discussed in Chapter 17, it allowed for the massive use of inexpensive, general-purpose processors instead of custom-produced supercomputing silicon. Falling prices for memory and hard drives helped too [54].

In the Distribution, communications companies added powerful Internet routers, fiber-optic channels, and began deploying broadband in residential areas. As is typical during this system growth phase, a lot of overinvestment in infrastructure occurred and the Internet boom of the late 1990s turned into a dot-com bust of the early 2000s. Nevertheless, even after the bust, the web remained alive and well.

The number of web users grew rapidly and a new trade-off emerged. On the one hand, the technology allowed for creation of very sophisticated, interactive pages, with scripts, media, ads, and advanced security features. On the other hand, creating such pages required high technical skill or significant resources to hire technical experts for website development. Compared to other popular Internet applications, like chat, email, and Usenet groups, interacting on the web or making frequent updates to your own pages was becoming taxing for an average, nontechnical user. The problem set the stage for the next phase of web evolution.

20 The Web Is Dead
Abandoning Documents and Files in Favor of Information Streams

The unconscious treats information like a fluid, not a solid.

—**David Brooks**, *The Social Animal* [55]

Today, what can be easier than tweeting a sentence or two? With a few keystrokes, one can show and tell the whole world what he or she is up to. On Facebook, adding pictures and a short note takes just seconds. With more than 800 million users and 130 friends per user on average [56],* messages ripple through billions of virtual channels, reaching people through web, email, and mobile applications. Facebook and other social networking companies now work relentlessly to speed up the pace of interaction. For example, early trials show that a simple increase in the number of bookmarks on the top of your page from four to six leads to a 20 percent increase in referrals from the bookmarks to games. More interactions means greater profits, either through ad insertion or increased spending of virtual currencies. Liberated by the technology, information pours in [57].

Often, this incredible growth in online activity is lumped together under the label of Web 2.0. Though definitions vary, the main driver behind Web 2.0 is considered to be user-generated content (UGC). That is, in contrast with the old web, the new web lets people create content through social networking, blogging, wikis, and other interfaces that do not require technical skills. What the Web 2.0 label obscures is the simple fact that the most popular social networking services like Facebook and Twitter don't need the web at all. Instead of the browser, they work great through dedicated applications on smartphones and tablets. Even if the web completely disappeared as a result of some technological disaster, Facebook and Twitter would survive the catastrophe. On the other hand, Wikipedia, despite also

* Dunbar's number is a suggested cognitive limit to the number of people with whom one can maintain stable social relationships. No precise value has been proposed for Dunbar's number. It has been proposed to lie between 100 and 230, with a commonly used value of 150.

being called a Web 2.0 service, would have a much harder time coming back. Why? Because it is designed and built as a collection of linked *documents*, just as Tim Berners-Lee envisioned the web in the early 1980s. The new Internet services represent a radical departure from the web paradigm.

IS THE WEB DEAD?

In 2010, online magazine *Wired* published an article titled "The Web is Dead." Its authors, Chris Anderson and Michael Wolf, noticed that the web's share of the Internet traffic had been declining since 2000, being squeezed out by video and peer-to-peer networking (Figure 20.1). They argued that mobile devices would further shift information consumption away from the browser to applications, and cause major changes in programming languages and data encapsulation technologies. They also emphasized an accompanying transition in content-related business models from free or advertisement-supported pages to *freemium* apps, with free basic access and paid subscriptions or purchases for premium content. In conclusion they wrote, "The Internet is the real revolution, as important as electricity; what we do with it is still evolving" [58].

What is the direction of this evolving revolution? From a system perspective, we can see a wave of technological growth, which is going to become much bigger than the original web. The web is not going to die, but it will become just one part of a much larger system containing a wide variety

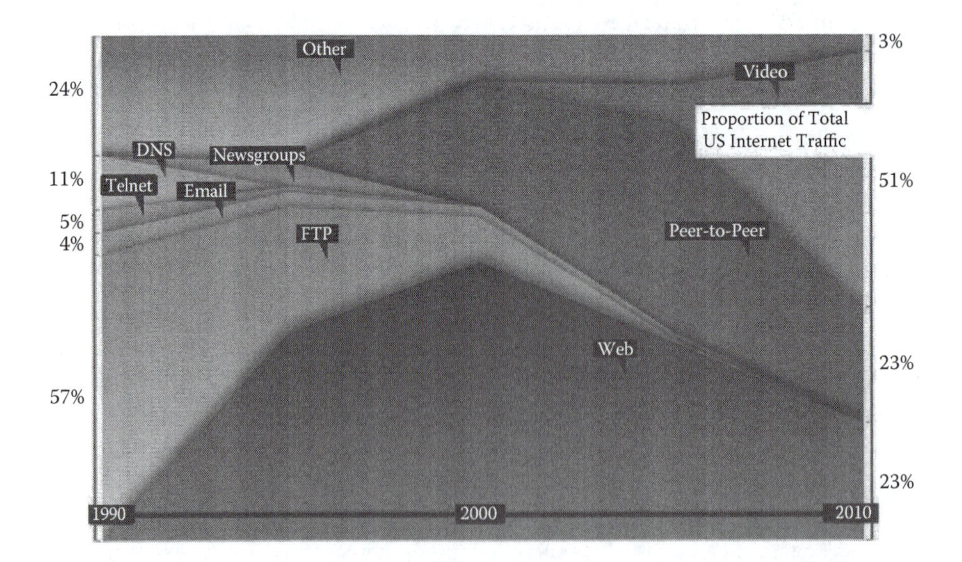

FIGURE 20.1 The web is dead. (From http://www.wired.com/magazine/2010/08/ff_webrip/all/1.)

of content sources and new modes of user interaction. Specifically, human interaction with information is shifting to mobile devices with touch-screen interfaces, and we are moving away from the document-and-mouse to a stream-and-zoom mode of accessing content.

EXAMPLE 1: WIKIPEDIA VERSUS FACEBOOK

To illustrate the nature of the technology transition, let us use the memory palace method we discussed briefly in Chapter 9. According to the method, to remember a large number of objects we create an imaginary building with many different rooms. Then we walk through the building, placing the objects into the rooms and remembering the path we used to get there. To recall the objects, we walk through the rooms and gather the objects. We will use the method to highlight structural differences between two of the most popular Web 2.0. services built with content generated by their users: Wikipedia and Facebook.

Wikipedia

We can think of Wikipedia as a memory palace where every room carries a tag with the name of an object. When we walk into the room we see pictures and text descriptions related to the object. Some of the descriptions have magic buttons next to certain highlighted words or phrases. The Wikipedia palace masters call these buttons *links*. By pressing any of these buttons, we jump instantaneously to a room tagged with the highlighted word or the phrase, and continue our exploration of the palace from there. The palace might remind us of a museum. Except when we feel that descriptions or pictures on the walls are wrong or incomplete, we are allowed to change them, or add our own pictures, descriptions, or magic buttons. We are also allowed to create new rooms and link them to other rooms by placing our magic buttons there. Such volunteer efforts have made the Wikipedia palace a practically infinite space where one can find rooms dedicated to almost anything in the world.

The palace is financed by donations and maintained by an army of volunteers from different countries. They expand the palace grounds (often called *servers*) and do a reasonably good job of removing garbage, intentionally or unintentionally left in the palace by visitors. Before Wikipedia, it was widely considered that only an elite group of paid professionals could create a high-quality memory palace experience called an *encyclopedia*. The professionals would build it out of paper or special memory disks and make money by charging visitors admission fees. The larger the palace, the more rooms with descriptions it houses, the higher the price of admission. In sharp contrast, the infinite Wikipedia palace is free to visit.

It is built with a low-cost technology called *the web*. Moreover, because it is being managed by so many volunteers from around the world, it has a lot more content: rooms, pictures, descriptions, and magic buttons. A popular web giant called Google, who people ask where to go when they want to see things, keeps referring them to Wikipedia. It is no wonder then that the old, venerable, paper-based palaces, like the *Encyclopedia Britannica*, are closing their doors.

The builders of the Wikipedia palace expect visitors not to buy things or play in the museum, but to study, observe, and help verify its exhibits. The builders strive for objectivity and truth. Therefore, whether one is a first grader or a professor, she often finds the same objects and the same rooms. The user interaction experience for everybody is also the same: generally, one is expected to read rooms/documents and navigate between them by pressing magic buttons/links with a device called the *mouse*. There's not much traffic between the rooms either. The palace as a whole is built as a collection of documents with links, thus fulfilling the vision of the web created in the beginning of the 1980s by Tim Berners-Lee. Overall, Wikipedia is always full of people, and despite the rapid growth in visitors who use devices with the new multitouch interface, there seems to be no urgent need for technology innovation there.

Facebook

Now, let us compare the Wikipedia palace with the Facebook palace. Though Facebook was originally created as a gated community, nowadays everybody is invited to build a room there. Most often, the room carries the name of its builder and contains objects that tell the life story of their owner and his or her friends. There is a constant stream of information between the rooms. As soon as one of the friends brings something or somebody into her room, says something, sees something, posts pictures or videos about something, or changes something in her life, all of her friends' rooms are updated. The friends are expected to comment on the events, "like" them, and spread information about them throughout the palace. An update on Wikipedia is just one event. Because an average Facebook dweller has 130 contacts, an update on Facebook generates a stream of 130 events sent to different rooms. In addition to that, people are encouraged to create common rooms, where they can discuss, create, and manage events together, from birthday parties to dancing flash mobs to revolutions.

There's good evidence that a reclusive Mr. Z, the founder of the Facebook memory palace, envisioned it as a platform for building a special brand of interactive rooms, called *social apps*. That was the big difference between his creation and MySpace, another popular memory palace built in 2003. Mr. Z gave app developers the ability to access people's life stories and their social connections, so that new experiences can be created and new events generated, thus increasing the intensity of information flows between the rooms. One of the most popular experiences in the Facebook palace is gaming. It is mostly free for the players, but certain game objects do cost money. To pay for them, Mr. Z created a local currency called Facebook credits, which the players can buy from Mr. Z with funds earned outside of the palace. When a player buys something during the game, the game developer and Mr. Z share the revenue.

Besides selling Facebook credits, Mr. Z and his crew make money by selling valuable advertisement space. The ads can be about objects or services sold inside as well as outside of Facebook. The more information people disclose about themselves, the more often they visit and interact with others in the palace, the more enticing are the ads, and the more precisely and more frequently they can be placed. Because for many, the objects and events in Facebook rooms comprise a status symbol, the Facebook palace is considered to be a great place to advertise objects, services, and experiences.

Unlike the Wikipedia palace where everybody can visit any room and see everything that's inside, Facebook palace dwellers have the ability to determine who has access to their rooms, their objects, their life stories, and their events. Accordingly, information streams in the palace have to follow access rules. Composing the streams, directing them, inserting ads, creating opportunities for people to talk and do things inside and outside of Facebook are all-important tasks for Mr. Z and his crew. Given that the number of people on Facebook is approaching one billion and connections between them are increasing as well, presenting rooms full of objects and conversations in a manageable fashion for any particular visitor is becoming challenging.

That is, the rooms are becoming crowded and the more recent events push out one's past experiences beyond the memory horizon, excluding them from potential conversations. To address the problem, Facebook introduced a feature called Timeline. It helps owners and visitors expand the rooms, enables people to zoom out of the present and then zoom onto a specific point in the past, be it a photo album, a conversation, or an event. Timeline is also a good advertisement analytics tool because knowing people's past choices allows advertisers and app developers to fine-tune their offerings to the tastes of a particular group of Facebook dwellers. Google, the web giant who loves open Wikipedia, has limited access to the Facebook palace. Because Google competes with Mr. Z for advertisers' money, Google is seriously concerned about the direction of the web and its technology. Maybe it is the reason why one of Google's leaders, Mr. B, said "I am more worried than I have been in the past. It's scary" [59].

21 Deconstructing Luck

Factors Affecting the Success of a System

Now that we understand at least the basic operational principles behind both Wikipedia and Facebook, we can consider how "lucky" they might become with the advent of new user interaction technologies.

In the previous chapter, we noted that the new multitouch, gesture-based interface technology doesn't provide any immediate advantages to Wikipedia. Is it any different for Facebook? The answer seems to be *yes*, because at least the zoom-in and -out gesture, already popular on multi-touch devices, is a good fit for navigating crowded Facebook memory rooms and timelines. Although any given room on Facebook or on Wikipedia is often denoted by the same word "page," Facebook pages are constructed as applications. They are designed to increase user interaction, generate, and receive streams of updates. As we will see (and can already anticipate from experience), the new multidirectional scrolling feature of the new touch-based user interaction paradigm should favor Facebook pages much more than Wikipedia pages with respect to encouraging interaction with such applications and streams.*

Moreover, once the Facebook mobile app and other Facebook-based apps are free from the constraints of the web browser, they should be able to introduce new modes of user interaction and improve the user experience. Although both Wikipedia and Facebook are often referred to as Web 2.0 applications hosting user-generated content (UGC), we can see that they represent two different technology systems. One is based on the old *document-and-mouse* GUI originated at Xerox PARC, while the other seems to transition to the new *stream-and-zoom* dominant design introduced by Apple with their iPhone, iPod, and iPad line of devices. The user interface aspect of this design is sometimes called zoomable user interface (ZUI).

* By the end of 2011, the total historical number of edits in Wikimedia projects, including Wikipedia, WikiNews, Wiktionary, and others, was about 1.5 billion (http://toolserver.org/~emijrp/wikimediacounter/). This was approximately one half of daily user "likes" and comments on Facebook (http://www.technologyreview.com/article/427678/facebooks-timeline/).

The emergence of a dominant design, as we noted in Chapter 13 about technology shifts, should be considered an important milestone in the process of system evolution because it points to a change in the direction of innovation efforts. Moreover, the emergence of an interface-related dominant design tells us about the accelerating pace of innovation afoot on both sides of the interface. For example, the GUI-based mouse-over-document PC interface led to a rapid growth of software applications, which brought about new computer experiences for users, from word processors to spreadsheets to games to the web. It also accelerated the development of novel PC hardware, peripherals, and networking. At the time of writing of this book, the mobile computing industry appears to have adopted the new multitouch user interface for the growing category of smartphones and tablets. What should we expect from this shift? How would these developments affect the electronic book distribution system many years down the road when it enters the Efficiency stage?

We have already seen an incredible growth in the number and variety of applications that target mobile devices. For example, Apple's App Store claims to have 500,000 apps in more than a dozen categories. Similar trends can be observed in competing Android-based application marketplaces. While most of us thought that the computer mouse was one of the simplest devices to operate, it appears that the multitouch screen is even simpler than the mouse. That is, we can now routinely see children barely out of their toddler years playing with their parents' iPads and iPhones. Zooming in and out of pictures, flipping, scrolling, shaking, and rotating the device brings out totally new, but intuitively understood interactive experiences.

We can also see rapid hardware developments for mobile devices, which bring to users displays with higher and higher resolutions, more memory and storage, faster processors and network access, new cameras, sensors, and other features. Though unsubsidized hardware costs still seem to be too high for massive adoption in the consumer and enterprise markets, industry projections estimate the number of connected mobile devices at 10 billion units, including 2.5 billion touch-screen tablets [60].

So far, in system terms, the discussion has involved technology changes related mostly to instances of the Tool. But as our discussion of Facebook versus Wikipedia shows, the Packaged Payload is moving toward stream-based implementations. In addition to Facebook, tens and hundreds of millions of users of other popular social networking environments, such as Twitter, YouTube, Instagram, Zynga, Pinterest, and Renren generate billions of real-time interactions. As the virtual networked cities grow, the pace of information exchange increases as

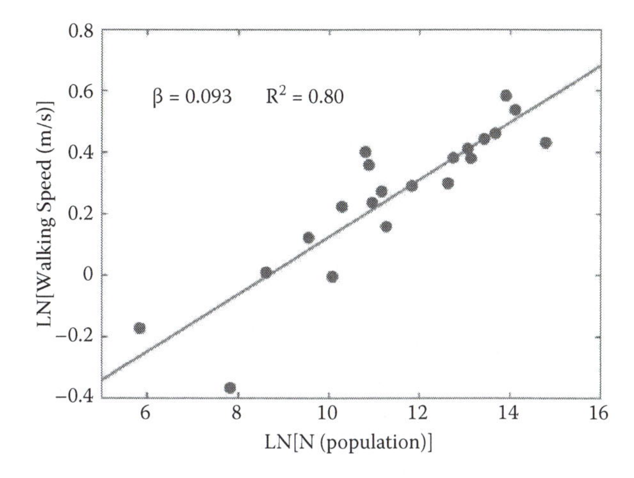

FIGURE 21.1 Walking speed as a function of city population size. (From http://www.pnas.org/content/104/17/7301.)

well. This is similar to the scalability pattern discovered in physical cities. That is, urban environments represent scaled versions of each other, where the logarithm of the size of the population in a metropolitan area is directly proportional to the logarithms of wages, number of supercreative people in the area, number of patents, crime level, and other social dynamic characteristics [61] (Figure 21.1). In a telling example, the larger the city, the faster people walk on its streets. In other words, an increase in city size leads to an increase in the pace of city life (Table 21.1).

Unlike physical cities, where urban environment changes organically, social networking environments have the ability to shape user experience proactively. For example, due to their transaction and advertisement-based business models, companies strategically design opportunities for short, recurring interactions that generate social events and the associated revenue generation opportunities. According to the *MIT Technology Review* [62]:

> [A] Zynga game generally asks players to perform quick activities: click here to plow a field in FarmVille; click here to fight a rival in Mafia Wars.
>
> The games are also meant to be conversation starters: you are encouraged to invite your Facebook friends to play with you and team up on various tasks, though you don't all have to be online at the same time for it to work. At nearly any given time, if you stop playing, it's easy to pick up where you left off.

TABLE 21.1
City Size and Urban Life Characteristics

Y	β	95% CI	Adj-R^2	Observations	Country, Year
New patents	1.27	[1.25, 1.29]	0.72	331	U.S. 2001
Inventors	1.25	[1.22, 1.27]	0.76	331	U.S. 2001
Private R&D employment	1.34	[1.29, 1.39]	0.92	266	U.S. 2002
Supercreative employment	1.15	[1.11, 1.18]	0.89	287	U.S. 2003
R&D establishments	1.19	[1.14, 1.22]	0.77	287	U.S. 1997
R&D employment	1.26	[1.18, 1.43]	0.93	295	China 2002
Total wages	1.12	[1.09, 1.13]	0.96	361	U.S. 2002
Total bank deposits	1.08	[1.03, 1.11]	0.91	267	U.S. 1996
GDP	1.15	[1.06, 1.23]	0.96	295	China 2002
GDP	1.26	[1.09, 1.46]	0.64	196	EU 1999-2003
GDP	1.13	[1.03, 1.23]	0.94	37	Germany 2003
Total electrical consumption	1.07	[1.03, 1.11]	0.88	392	Germany 2002
New AIDS cases	1.23	[1.18, 1.29]	0.76	93	U.S. 2002–2003
Serious crimes	1.16	[1.11, 1.18]	0.89	287	U.S. 2003
Total housing	1.00	[0.99, 1.01]	0.99	316	U.S. 1990
Total employment	1.01	[0.99, 1.02]	0.98	331	U.S. 2001
Household electrical consumption	1.00	[0.94, 1.06]	0.88	377	Germany 2002
Household electrical consumption	1.05	[0.89, 1.22]	0.91	295	China 2002
Household water consumption	1.01	[0.89, 1.11]	0.96	295	China 2002
Gasoline stations	0.77	[0.74, 0.81]	0.93	318	U.S. 2001
Gasoline sales	0.79	[0.73, 0.80]	0.94	318	U.S. 2001
Length of electrical cables	0.87	[0.82, 0.92]	0.75	380	Germany 2002
Road surface	0.83	[0.74, 0.92]	0.87	29	Germany 2002

Source: Luís A. Bettencourt et al., Growth, innovation, scaling, and the pace of life in cities, *Proceedings of the National Academy of Sciences* 104, no. 17 (2007), http://www.pnas.org/content/104/17/7301.

Another important development is the increasing weight of streams that flow through the system.

By *weight* we mean the amount of raw data included in the stream. In 2010, Internet video took the largest share of network bandwidth. According to Dr. Regina Dugan, the director of DARPA, "More video is uploaded in 60 days than has been created in 60 years by 3 major US TV networks combined" [63]. Short video clips shared and streamed from YouTube and other sites are extremely popular among users of mobile apps [64] (Figure 21.2).

It is reasonable to predict that as mobiles get more storage, better cameras, and greater access to high-bandwidth networks, social media services

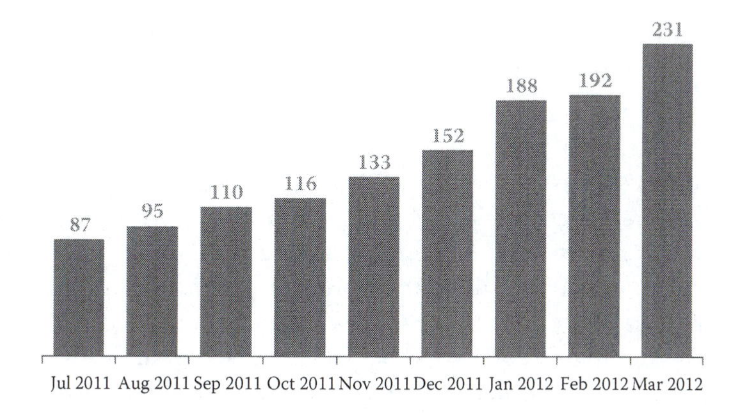

FIGURE 21.2 Monthly minutes per active user, photo and video apps. (From http://blog.flurry.com/.)

like Instagram will provide even more opportunities for real-time sharing of user-generated videos. In another example, Google is testing its augmented reality glasses, capable of two-way streaming video. According to the *New York Times*:

> The glasses can stream information to the lenses and allow the wearer to send and receive messages through voice commands. There is also a built-in camera to record video and take pictures [65].

As we discussed earlier, in contrast with document-based Packaged Payloads, the new composite streams of information create additional opportunities for multitouch user interface applications, because they contain multiple layers of information that can be experienced within different space–time frames.

The shift from documents to streams also entails a change in Source implementations. One reason for this is the increase in granularity of the Packaged Payload that has to be assembled by the Source. That is, as we discussed earlier, the streams may contain more information pieces, which come in a greater number of different sizes. Because of the increased intensity of user interaction, the streams have to be assembled faster and customized to a particular user experience. This development diminishes the viability or interactivity of the more traditional broadcast-type information.

For example, let us compare assembling a page, that is, a memory room, for Wikipedia and Facebook, respectively. The former contains a mostly fixed set of pieces, such as text, pictures, and links. Further, the page is the same for all users, which means that once assembled for the first user, its cached copy can be kept in memory to serve other users.

In contrast, a Facebook page, or more precisely a stream feeding a Facebook app, has to be put together from status updates, game events, pictures, tweets, likes, inserted videos, friends' comments, ads, scripts, and many more. Since access rights can be different for different users, the Source has to apply different rules during the process of the stream composition. Moreover, similar but different streams have to be assembled for third-party applications that use Facebook's social graph and timeline information. Compared to Wikipedia, Facebook's stream "packing" logic has to be much smarter, faster, and more precise. For instance, a wrong snippet of text on a Wikipedia page would be a factual mistake; users can correct it easily. On the other hand, a Facebook status update sent to a wrong user would be a privacy violation, social and business consequences of which sometimes could not be undone.

The intelligence of Facebook's internal implementation of the Source functionality is essential to the company's success, as it grows by adding more users, applications, content, and relationships to their social network. It should be appreciated that the Source can be thought of as a subsystem of a larger system. As we discussed in Chapter 3 about plasmons and trains, the subsystem comprises its own Tools, Sources, Packaged Payloads, Distributions, and Controls. Its smarts represent the Control functionality, which is essential for Facebook's success. The company's patents, coauthored by Aaron Sittig and Mark Zuckerberg, were originally filed right before Christmas of 2005. They show the inventors' focus on the Control as a critical part of what they call the Social Network Engine (Figure 21.3).

Both patents carry the title of "Managing Information about Relationships in a Social Network via a Social Timeline," and cover different aspects of assembling and presenting various pieces of user data according to the rules determined by social connections and privacy preferences. Like in any other system, a large number of fast-moving parts inside the engine create the need for a sophisticated management component, the Control function, which Facebook has decided to protect with patents from the early days of the startup.

Let us get back to the higher-level system and use the Facebook patent diagram to illustrate the changes in the Distribution, necessary for a successful transition to stream-based interactions. The diagram shows it as a cloud tagged *network*, the cloud magically connects the engine to a plurality (a number greater than two in legal jargon and almost a billion in real life) of users. How does it handle all these connections? What kind of problems do the information streams create for the network fabric?

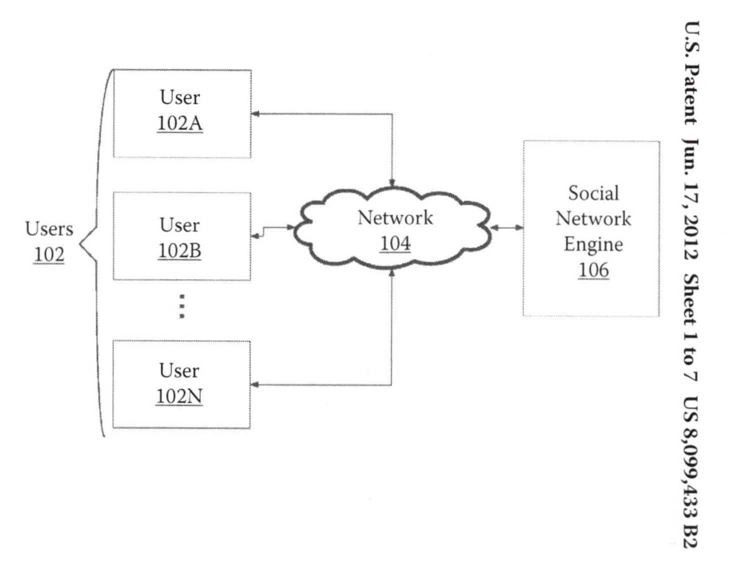

FIGURE 21.3 Aaron Sittig and Mark Zuckerberg, US Patent 8,099,433.

The evolution of the Internet as a networking architecture is outside the scope of this book. But for our purposes, we do need to know that ideas about the Internet originated during the Cold War years as a command-and-control network capable of surviving a limited nuclear war between the United States and the Soviet Union. One of the war scenarios involved a complete destruction of key nodes on the network (Figure 21.4a and b). The architecture of the telephone or radio network, which was the main communications medium at the time, would fail under this scenario because it relied on its central offices. The offices were responsible for making connections between incoming and outgoing communication lines. In a greatly simplified example, if a central office handling the 415 area code was destroyed, the city of San Francisco and its vicinity would be completely cut off from the rest of the country.

In the period between 1960 and 1962, Paul Baran (1926–2011), then an engineer at the RAND Corporation, proposed a series of solutions for the network problem. He suggested that a reliable, high-speed digital network could be built using unreliable links and nodes. That is, when we change the network's layout from centralized or partially centralized to distributed, it becomes capable of routing data around destroyed links and nodes. But changing the layout, or the Distribution in system terms, would not be enough. We would have to synthesize a new type of system with different implementations of the Packaged Payload, Control, Source, and Tool.

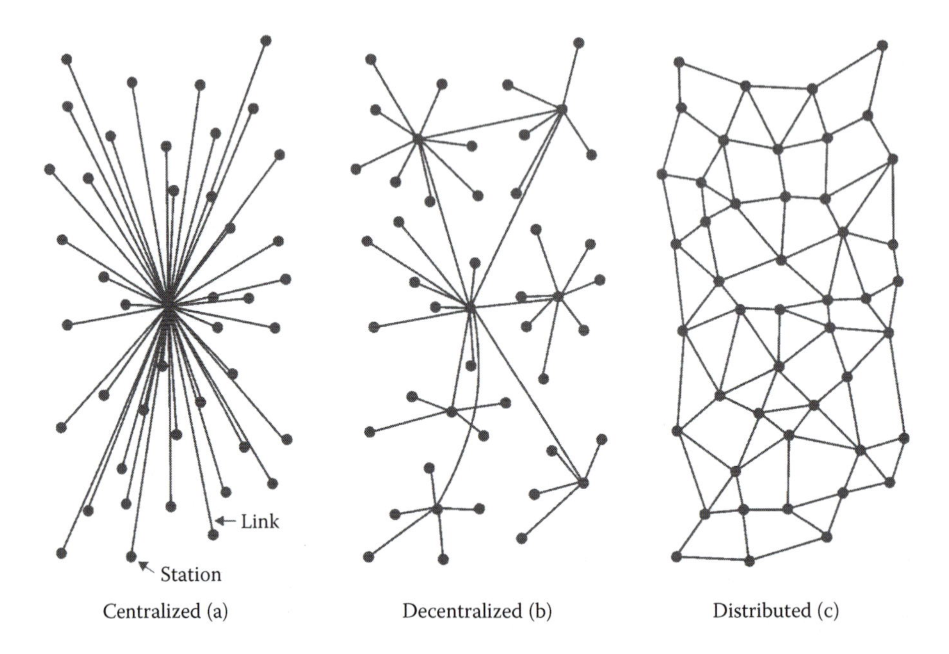

FIGURE 21.4 Network architecture diagrams from Baran's 1962 paper on network reliability. Diagrams courtesy Rand Corporation [66].

One of Baran's key ideas was to split the message (the Packaged Payload) into multiple chunks or message blocks at the origin (the Source), and assemble the message from the blocks at the destination (the Tool). Why the complication? Why can't we send the message in its original form? Because when the network fails to deliver the message, the entire message has to be resent. For instance, if a network segment carrying the message is destroyed, the message is lost, and we might never find out about it. Similarly, when the message is lost due to a routing failure, this mishap has to be discovered and the message resent. On the other hand, if we send out many, sometimes duplicate message blocks, even if only one of them goes through, the destination node can request the origin to resend the rest. Although this approach may cause an increase in message delivery time and generate some extra traffic, as long as there exists at least one network route *alive* at any given time, eventually the entire message will be delivered successfully.

Note that to handle the increase in number and variety of nodes, message blocks, routes, and other network elements, the system has to become smarter. It has to learn how to split and assemble messages, deal with failures, communicate with nodes, set delivery policies, optimize network layout, handle message priorities, and so on. In other words, the Control functionality has to be reinvented and reimplemented too.

As we discussed earlier in the book, solving the synthesis problem is probably the most difficult task to accomplish in the process of system evolution. It took many years and a lot of human ingenuity to further flesh out the original ideas and implement the network architecture we now call the Internet. Although the nuclear war scenarios—thank God—never materialized, building a reliable data communications system out of unreliable parts was a promising concept. At the time, both computer hardware and software were disparate and unreliable; communication equipment and protocols, intermediate network nodes, and links between them were not too reliable either. Amazingly, despite all these problems, by mid-1970 a message sent from Stanford, California, to Cambridge, Massachusetts, using the Advanced Research Projects Agency Network (ARPANET) went through.

In 1960, when Paul Baran envisioned a future system for digital communications, the standard speed of text transmission using a teletype channel was 60 words per minute. By comparison, the Google Fiber project promises to bring residential network bandwidth up to 1 gigabit per second, which is about 30 million times faster than the teletype. But it's not only the available network bandwidth that is growing. The variety of data that flows over the network is increasing as well. We send emails, videos, pictures, web pages, tweets, links, and other forms of information.

As we discussed earlier, video has become the dominant form of information delivered through the Internet. Unlike emails or web pages, video streams are much more time and loss sensitive. If a certain video frame is lost or delayed, an entire video sequence cannot be played. Therefore, sending and receiving video streams requires not only splitting a movie into billions of data packets, but also making sure that the packets are given priority over other kinds of data. In another example, if the network fabric loses a high-frequency stock trade request, resending it even a second or two later no longer makes business sense. To summarize, in addition to increasing the bandwidth of the Internet (for instance, by laying fiber-optic cables across the world), we also have to make the network fabric robust and smart.

This is the kind of problem that Silicon Valley startups like Nicira help solve for AT&T and other telecom giants. Founded by researchers from Stanford University and the University of California Berkeley, the startup developed a software platform for virtualizing the network fabric (Figure 21.5).

Their solution provides a stable interface between the software and the hardware that compose the network's routing intelligence. As we learned earlier, the emergence of an interface enables rapid technology innovation on

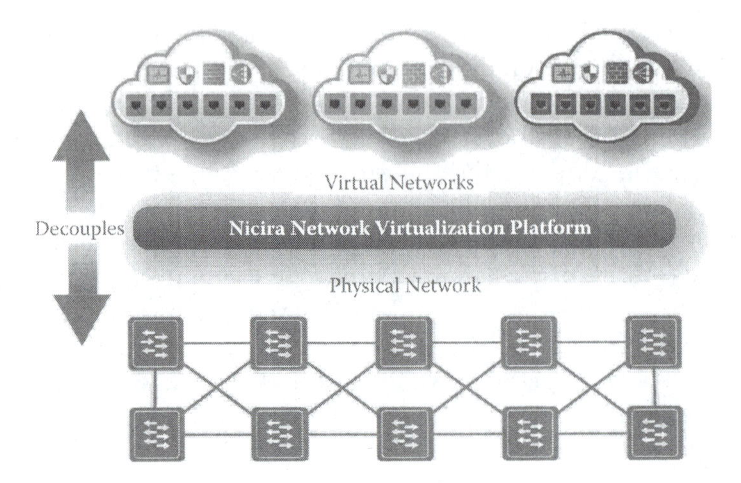

FIGURE 21.5 Nicira platform as an interface between network software and hardware. The interface layer separates hardware and software, enabling rapid pace of innovation for both. Diagram courtesy of Nicira. US Patent Application 20100257263.

both sides of the interface. In Nicira's case, the interface allows network hardware to become faster, while software data processing algorithms become more flexible. As a result, the network improves its handling of the growing volume and variety of information streams. In system terms, the network shown on the Facebook patent diagram (Figure 21.3, Pos 104) provides the Distribution functionality, while the routing software and hardware play the role of the Control inside it. The efficiency of the Distribution depends critically on the intelligence of the Control to route trillions of packets of data to their destinations in a timely fashion.

While identifying the Control functionality within the Distribution component is quite easy; finding instances of the Control component at the Internet layer is a more difficult task, which we will consider next.

REVIEW QUESTIONS

1. Develop a user scenario where a Google Glass device helps people solve the worldwide obesity epidemic discussed in Review Questions after Chapter 12. Which system element(s) would the device implement? Sketch out the rest of the system. What elements are missing? Describe a high-level industry shift that would make your scenario very successful (review Chapter 15 if necessary).

2. Let us assume that the web is dead. Name at least two currently dominant designs that will disappear along with the web. Explain your choices.

22 Seeing the Invisible

The System behind the New Internet

The diagram on Figure 21.4 seems to confirm the idea that Facebook is all about user-generated content (UGC). That is, the system works when Internet users send their content, such as pictures, videos, status updates, and messages, to Facebook's Social Network Engine 106 via Network 104. Though the existence of the content (i.e., the Packaged Payload that encompasses user-generated and user-derived data on Facebook) is implied, the Control functionality is not in the picture at all. Where is it, then? Or maybe, it simply doesn't exist?

One of the classical problems in representing a situation with a drawing is that it is not always possible to draw the picture and show that a specific item is *not* there. For example, it is possible to draw a picture of a room with no elephant in it. Does it show that the elephant is not in the room or that the elephant is there, but it is hiding in the closet? Furthermore, how do you distinguish between a drawing of a room with no elephant from another drawing of the same room with no rhinoceros?

A common way to show a missing elephant in the room is to hint at the animal's past, current, or future presence there. We would show the tip of the elephant's trunk sticking out of the closet, a chance reflection in a closet mirror, a cage with an "Elephant" sign on it, and so on. Jigsaw puzzles solve the missing elephant problem by showing a vacant space where the elephant or the rhinoceros is supposed to be. But to see the vacancy, we need to know already that we are dealing with a jigsaw puzzle and have other pieces of the puzzle fall into their proper places. In this regard, models—for example, mathematical or biological ones—help us create puzzles that mimic a real situation and identify missing pieces. That is why, according to philosopher of science Thomas Kuhn, in contrast with paradigm shifts, "normal science" often can be considered a puzzle-solving process. Unfortunately, puzzles rarely help us see what lies outside of them. To see beyond puzzles, we need to think outside the box of our "drawings" of reality.

Bearing this in mind, let us continue exploring the Facebook picture. We have already established that instances of the Packaged Payload are missing.

Although one could argue whether the Control functionality is present or not, it is relatively easy to recognize that the time dimension of user interaction with Facebook is barely present in the original patent diagram. That is, the picture, though very useful in focusing our attention on the existence of users, the network, and the engine, provides a single snapshot in time. Luckily, the four unnumbered two-way arrows give a hint that there's more to the picture. What do the arrows represent?

Each arrow shows the direction of information flow between the users and Facebook. For example, one step in the process would be when User 102A sends her photo to Social Network Engine 106. Another step would be when the engine forwards the photo to her friend, User 102N. Also, we can easily imagine that there is a step in between, when the engine stores the photo. Having redrawn (Figure 22.1) the original picture to reflect the timing aspect of the transaction, we can see that in addition to the Source functionality discussed in Chapter 20, the Wikipedia versus Facebook chapter of this book, Social Network Engine 104 also has the Distribution functionality at the higher system layer, where Facebook is one of many Internet services. Other services include Google, Yahoo, Wikipedia, Amazon, and so on. In the photo-sharing example, the photo gets distributed from one user to another exclusively through the Facebook engine.

Note that just like in the missing elephant picture, we can't see users who don't receive the photo, due to the fact that they are not friends with the first user. In other words, the social networking engine sees everything

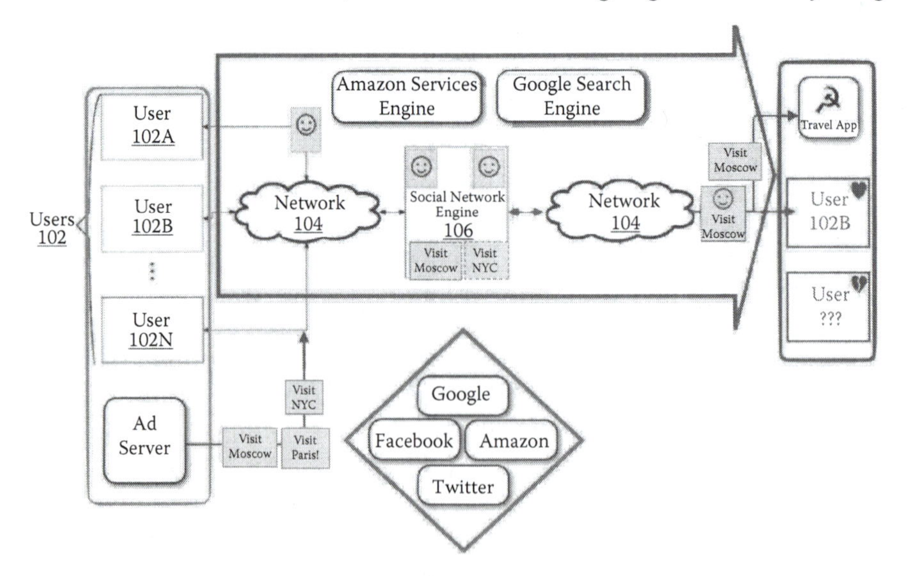

FIGURE 22.1 A higher-level view of Facebook and other Internet services.

users distribute through Facebook, but the users see only what their friends, knowingly or unknowingly, instruct the engine to share with others. In fact, the users delegate execution of the Control functionality to the engine. Furthermore, because Facebook sets privacy defaults, defines sharing mechanisms, and designs user interfaces, the company has more control over information sharing than many users realize.

For example, in 2006, when News Feed, a feature that automatically propagated events to friends' pages, was introduced, it was met with surprise and disbelief regarding how much of a user's personal information that user's friends could see. Despite initial complaints, the feature became a mainstay of the Facebook experience because it allowed people to stay in touch automatically. In 2007, Facebook launched its Beacon service, which allowed external websites to send automatic status updates to Facebook members without their explicit approval. In combination with News Feed, the updates would propagate further to one's Facebook friends. For example, when someone would buy a present or a condom, all his Facebook friends would immediately learn about both purchases. After much controversy, including a class-action lawsuit, the service was eventually shut down in 2009.

When we add another invisible elephant—Facebook apps—to the picture, we find that they can access user information in a manner neither the users nor Facebook anticipated. In 2010, the *Wall Street Journal* discovered that apps "have been transmitting identifying information—in effect, providing access to people's names and, in some cases, their friends' names—to dozens of advertising and Internet tracking companies" [67] In 2012, the *Journal* investigated the 100 most popular apps and found that they routinely acquire and provide users' personal information, such as email addresses, location, sexual preferences, and so on to third parties, including advertisers who were not approved by Facebook [68].

Though Facebook provides users with a wide range of options to control the flow of personal information, the overwhelming majority of users don't exercise these options because this would require a conscious sustained effort. Moreover, the users frequently have erroneous perceptions about their own level of online sophistication. In a study of 103 participants (61 male, 43 female), practically all participants used just two levels of privacy—*everyone* and *friends-only*—while most of them had the perception they were using four levels—*everyone, friends-only, friends of friends,* and *customized.* (Figure 22.2).

Although females expressed more concern about their privacy, in reality they posted more often and with fewer restrictions than males [69]. In addition to this perception–reality gap, the 2012 *Wall Street Journal* study showed

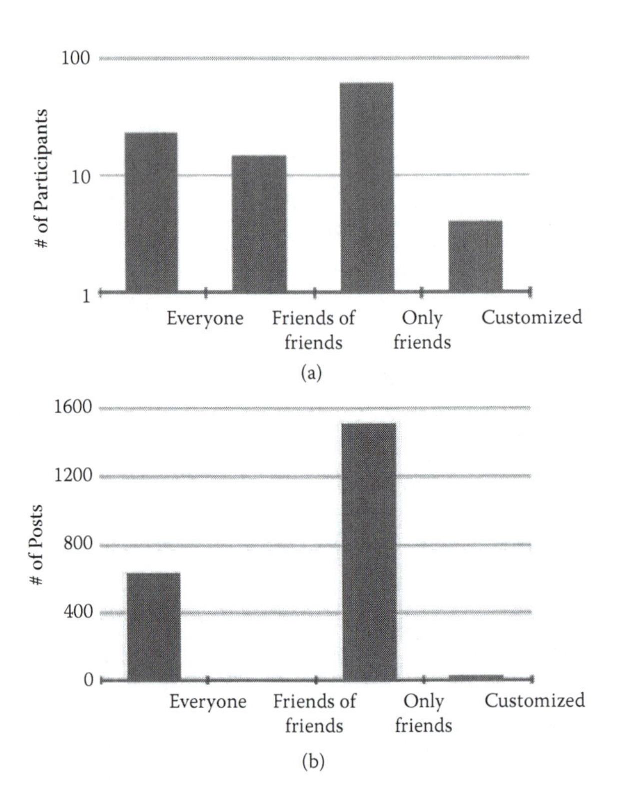

FIGURE 22.2 (a) Stated versus (b) actual distribution of privacy settings for individual posts.

that Facebook friends disclose somebody else's personal information without thinking too much about it. As a result, users delegate to Facebook more Control functionality than they are aware of.

Let us emphasize that the purpose of this discussion is not to judge the quality of Facebook's privacy policies, either stated or implemented. However, with regard to the properties and importance of the Control functionality of a given system, further discussion is illuminating.

Within a broader context of data security breaches at organizations whose primary function is to secure sensitive data, Facebook looks remarkably good. To compare, in the 2009 Wikileaks scandal, the US government lost over 260,000 classified diplomatic and military documents, allegedly due to the actions of United States Army Specialist Bradley Manning. The veracity or falseness of these particular allegations doesn't preclude the real possibility of a security system being prone to leaks. In another example, in 2011 it was disclosed that hackers stole authentication secrets from RSA, a company characterizing itself as "the premier provider of security, risk and compliance management solutions for business acceleration." As a result, tens of millions of SecureID tokens used, among others, by employees at major US government contractors and global banking firms to log onto corporate networks, had to be replaced. The full extent of confidential data "lost"* at those organizations because of the SecuredID attack is unknown.

With regard to personal data, millions of users lose it—or more accurately, lose the full control of it—through hacker attacks on financial institutions, government agencies, and popular Internet services. To mention just a few of them in the order of scale, according to datalossdb.org, an estimated 130 million user records were compromised or stolen at Heartland Payment Systems, a credit card payment processing company, in 2009. In 2011, hackers attacked Sony Playstation Network and stole data related to 77 million individuals; names, email addresses, birthdates, passwords and logins, purchase history, and possibly credit cards. In 2006, names, Social Security numbers, and other personal information of 26 million people were stolen from the US Department of Veterans Affairs. In 2012, Global Payments Inc., a self-described "leader in payment processing services," reported that "less than 1.5 million" credit card numbers were "exported" in an unauthorized transaction [70].

In view of this situation, the security and user privacy management system at Facebook appears to be quite robust. Moreover, the company's spam-fighting efforts are unparalleled in the world. With more than 25 billion transactions every day, only 1 percent of users experience spam [71]. Facebook, being by far the most popular social networking website on the Internet (Figure 22.3), has developed a sophisticated defense known as the Facebook Immune System (FIS).

A team of 30 people and a set of artificial intelligence (AI) algorithms oversee FIS, which in human society terms, is less than 0.004 policeman per 100,000 users [72]. In comparison, the United Nations recommends a minimum police strength of 222 police per 100,000 people [73]. The FIS is another element of the Control functionality Facebook provides for its users. That is, Facebook controls the flow of spam (unwanted content), so that it does not feed into user information streams. With the magnifying effect of Facebook's event propagation mechanism, allowing even small amounts of data pollution would create an informational disaster.

* The information wasn't really lost. In fact, it probably just got duplicated, and someone's loss is another's gain. The loss was that of control.

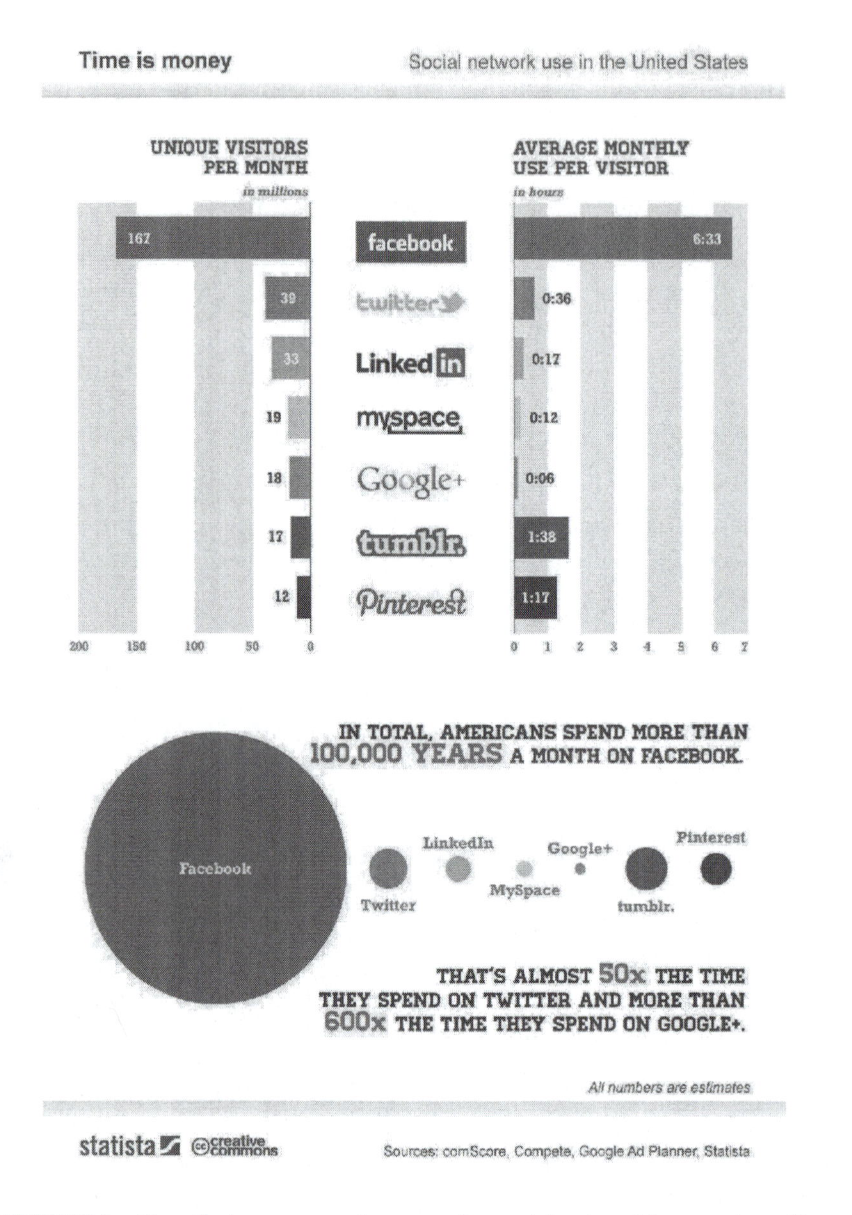

FIGURE 22.3 Time that users spend on popular social networking services. (From http://venturebeat.com/2012/02/17/facebook-engagement/.)

GOOGLE VERSUS FACEBOOK: THE BATTLE FOR THE CONTROL FUNCTIONALITY

Jim Larus, a researcher at Microsoft who studies large networks, estimates that the Facebook network is second only to the World Wide Web itself in connectivity. An important Control aspect of Facebook functionality felt across the Internet is the sharing of links by Facebook users and the amazingly simple, yet efficient *Like* button. Both of them go beyond the UGC label. These features (the ability to share and *like* or point a virtual finger at something) help users direct their friends' attention toward the content they consider remarkable, regardless of the location and time separating the users.

Before Facebook, Google was the dominant force in the business of recognizing what information people wanted to see on the web. Its web indexing engine would analyze the network of links and decide on page rankings, as they related to search terms entered by users. Google founders Larry Page and Sergey Brin developed their page ranking algorithms while they were computer science graduate students at Stanford University. Their key insight was that higher-quality pages were linked to other pages more often. That is, Page and Brin realized that web content developers expressed their content preferences by linking to the pages they liked. When a Google search produces a ranked list of links, it factors in the number and quality of references.

As we discussed earlier, to simplify the maintenance of a large collection of linked documents on the web, Tim Berners-Lee dropped the requirement for a two-way link between web pages. Therefore, a web page doesn't *know* about the web pages linking to it. This piece of information, crucial to Google's relevancy algorithms, has to be discovered independently. Ideally, to gain comprehensive data about the structure of links, one needs to examine the entire World Wide Web—a task Google took upon themselves from the very beginning. As applied to the web, their mission "to organize the world's information" is not a metaphor, but rather a specific technical task essential for producing high-quality page rankings for search, which is in turn essential to Google's business model. In addition to links, Google tracks and incorporates into the page rank information about user clicks, time spent on the page, and other statistics. In other words, Google directs user attention—an important element of the Control functionality—toward the pages its indexing and user tracking algorithms deem more valuable.

Relevancy-wise, Google's approach for organizing user attention is better than those of Yahoo's original web catalog method or document-based search techniques developed by AltaVista, Excite, and others during the 1990s. Nevertheless, when compared to Facebook, it suffers from at least three key problems that limit the robustness of the Control functionality on which Google's original business model relies.

First, Google's approach to organizing and revealing content is based on inferred preferences of an unknown (to the particular user of Google sending a request) group of people, rather than direct recommendations from that user's friends. Even though the "likes" are visible on the web page and can be counted by Google's indexer, it's hard to know whose likes they are and how relevant they are to a particular user. This limitation may compromise the ranking of search results and user satisfaction. In contrast, for Facebook, the task of collecting likes, which flow directly to it through the sharing mechanism, is technically simpler than indexing the whole web or tracking individual users across multiple websites.

Second, Google search is reactive rather than proactive. To be directed to a page, the user has to initiate the search. This reduces the opportunity to trigger ad views, impulse purchasing, and other forms of spontaneous user engagement. By contrast, on Facebook the user is presented with relevant information propagating in real time through his or her social graph.

Third, Google search works poorly with apps and information streams (like online games) because apps and streams generate user events instead of links. By contrast, Facebook enables a range of user activity, for example gaming, via its network, thus allowing the events to propagate through Facebook's social graph, not the web at large. So far, Google's attempts to retake the initiative using Google+ or the +1 button have produced limited results. Jon Stewart, the host of the *The Daily Show*, recently joked that if you don't want people to know what you've done, you should share it on Google+.

In March 2012, to help solve its Facebook problems, Google began tracking users across all its services and applications, including Search, Gmail, YouTube, Blogger, Analytics, Maps, Google+, the Chrome browser, ad preferences, Android apps, and others. With these steps (given Google's broad reach on the web and mobile services) the company is likely to maintain its strong position with regard to the system-level Control functionality. On the other hand, as long as people share their experiences on Facebook, Twitter, LinkedIn, or some other social environment outside of the Google realm, it makes little sense Googling how your best friend or most respected colleague feels about a particular book. Furthermore, Facebook's alliance with Microsoft allows the latter to compete with Google by providing search results, augmented with more complete social graph data. Although it is unlikely that Microsoft will unseat Google from its dominant search position on the web, the situation may evolve differently in the emerging world of information streams and mobile applications.

For example, imagine a library application from Amazon. It provides the reader with a stream of reviews, likes, tweets, and other reactions from fellow readers past and present. The stream is filtered using keywords, location, social graph information, and the context of the e-book being read. Searching for relevant content or its *aboutness* does not require the user to go to the web using a browser. Instead, the search is performed without the user leaving the library. Furthermore, social interactions, such as discussing books, making presentations, tutoring, playing a game, or even dating, happen inside the library app, rather than the web at large. When desired, events in the library are coordinated and reported through the social graph of the user's choice. To avoid distractions, the user or her authorized librarian can set the app in the proper study mode. In such an environment, orchestrating user experience on a large scale (implementing the Control), goes well beyond responding to search queries and providing relevant links.

Of course, one can imagine other digital library features, including 3D graphics and video conferencing, but the goal of this exercise is not to develop a detailed specification for a future app. Rather, its purpose is to help us discuss potential opportunities available to the users due to the evolution of the system when it reaches the Efficiency phase. Now that we have a more complete picture of an Internet-based system for information distribution, especially its higher-layer Packaged Payload and Control implementations that were missing from common representations, we are ready to develop long-term transformation scenarios that affect the future of such technologies, like the e-book.

23 The Book Electric
A Scenario of E-Book Evolution

Our previous examples of a system in its Efficiency stage involved configurations of the existing electric grid. Specifically, we considered the role of the Control functionality as well as systemwide disasters caused by Control-related failures. With the e-book we have to look into the future rather than the past. So, to illustrate the utility of system-level analysis for forward-looking projections, let us conduct a thought experiment in which we will apply systematic techniques discussed in previous chapters (the Three Magicians, Scale-Time-Money, etc.).

As we discussed earlier, the dominant design of the vast majority of contemporary books goes back to the codex, a two-thousand-year-old invention. That is, the book, including the e-book, looks like a collection of numbered pages with text and figures. In contrast with the scroll (an older writing technology), it enables direct access to each page. But what would a new dominant design for the *book object* look like?

Let us imagine it as a multidimensional interactive entity, something like a Google Earth app on steroids. We can play with it at will; use it for moving through space, time, and scale of events. We can explore Edison's innovations by observing the growth of railroads and the streams of coal they bring to cities and factories. We can zoom into a power station and see how workers shovel coal into a steam engine, turning the dynamo and producing electricity, which then flows through wires, meters, and switches all the way to the light bulbs. We can zoom in even further and observe how the light bulb works inside, or go back in time and visit Edison's lab in Menlo Park where he was working on his inventions. Staying at the same level, we can visit other inventors and compare their light bulbs to the one by Edison and appreciate the reasons his idea turned out to be the best. Everything we touch in this imagined world can be picked up and explored further, either through direct interaction or simply by reading the accompanying text.

We can move further in this new "book" by zooming back out and speeding up the clock to see how a new electric infrastructure grows. New factories with the connection to the grid are being built; stores, offices, and houses now have internal wiring because everybody knows that electricity is the way of the future. New electric appliances begin appearing in the house, a car shows up where horse carriages used to drive, a phone rings in the office, an air conditioner blows cool air in the theater. How did they get there?

If we are interested in technological details, we can explore the inner workings of a specific device. Alternatively, we may decide to follow the expansion of the highway system and car-friendly city streets, noticing the emergence of traffic lights, road signs, gas stations, and parking lots around the stores. Remarkably, only two or three degrees of technological separation lie between the world of railroads and coal and the world of interstate freeways, gas stations, and ubiquitous McDonald's restaurants.

If we are interested in business models, we may zoom into Edison's accounting department and see how profits and losses flow through time. By following the paper trail of his contracts, we discover what kind of deals the inventor had to make to become successful. Similarly, his patent filings and court actions can be traced and correlated with technology and business developments. Furthermore, the actions of his competitors, such as George Westinghouse and the Thomson-Houston Electric Company, are integrated into the information stream and mapped onto our interactive book world. By moving up to the level of concepts and then zooming back into other technical systems, we are able to explore and appreciate differences between various solutions, and identify dead ends and lucky turning points.

The experience of the book we are envisioning is gamelike. But rather than chasing an alien invader with a blaster, we learn how innovation unfolds in time and space. We may even invite other readers to explore it with us, ask questions, and provide answers. The book may offer different graphics or levels of technical detail, depending on whether the reader is a first-grader, an engineer, or a financial analyst. The experience can be both objective and subjective. That is, the factual information presented is objective, while the mode of presentation is customized to the reader's learning abilities.

Much of the information within this new book is already contained in other books, Wikipedia, and Internet-based archives. However, to achieve the functionality and utility described above, we need to create interfaces and coordinate the delivery of a drastically greater number of different Packaged Payloads carrying content for this system. If a large number of readers were to experience such an e-book at once, the sharing and availability of information must be vastly more flexible than what the conventional World Wide Web allows. The new e-Book system will have to be built as a slice of a higher layer of the Internet system in which

a. user devices (instances of the Tool) have rich, sensory, multitouch, multimodal interfaces;

b. servers (instances of the Source) store large amounts of information and respond quickly to user interactive requests;

c. the network fabric (instances of the Distribution) has the bandwidth and the smarts necessary for communicating user requests and delivering information in a timely manner;

d. the information itself (instances of the Packaged Payload) is constructed as zoomable interactive streams, comprising both original and user generated content; and

e. user setup and experience software (instances of the Control) orchestrate seamless interaction between all other system elements.

The new system looks drastically different when compared with today's book publishing and distribution industry. Nevertheless, it is consistent with the vector of change from the original document on a computer screen. The vector extends to the linked, unindexed World Wide Web of Tim Berners-Lee, to the indexed one in the age of Google, and then an information-intense, stream-based paradigm that is the backbone of Facebook, YouTube, and Zynga. The new e-book represents a library of interactive experiences and active learning, especially where educational topics are concerned. Furthermore, the system creates a unified, rather than compartmentalized view of the world by eliminating artificial subject boundaries imposed by centuries-old textbook printing traditions. For instance, when studying the brain, one can seamlessly navigate from the biological structure of the neuron to the physics of how the neuron propagates an electrical charge to the math of a neural network model to the medicine of human brain disorders. The e-book-as-a-library works by orchestrating various interactions with the material the user needs to not only read and comprehend, but also to practice and master.

BUILDING THE E-BOOK SYSTEM

Today, despite the abundance of relevant information, building the new "book" from our thought experiment would be a technically difficult and expensive enterprise. Typically, only professionals such as cinematographers, designers, and engineers have access to the complicated equipment for creating four-dimensional interactive environments. For example, using a specially invented video camera set, award-winning director James Cameron and his team created *Avatar*, a movie that lets viewers explore the amazing world of Pandora, an imaginary planet where human invaders collide with the locals over control of Pandora's natural resources.

Some of the objects in the movie were designed with 3-D modeling software tools similar to the ones used in modern engineering. In contrast with the traditional approach, instead of drafting cross sections and flat views of various machine parts, today's computer-aided design (CAD) software enables the engineer to model virtual 3-D parts first. After ensuring that the part fits into the rest of the machine, the engineer sends its electronic representation to the factory, where the part is produced. Some people believe that together with 3-D printing, this technology will enable revolutionary additive [74] manufacturing. According to the new method, rather than machining off extra material to produce the desired part (a *subtractive* process), the new system would allow for making the part by adding the material bit by bit or layer by layer. In any case, whether or not physical objects are manufactured by using subtractive or additive methods, it's important to realize that generating 3-D models of everyday objects is becoming a common engineering practice.*

Although creating a highly interactive fiction e-book using such 3-D models would still be a difficult task, producing a zoomable electronic user manual for automobile repair would be a relatively simple project. Just like the transition from paper (analog) to electronic (digital) documents was part of a revolution in computing and communications, modern 3-D design tools combined with the new Internet technologies have the potential to change the nature of our work with technical documentation. Of course, before this possibility is realized on a large scale, we would have to solve—speaking in system terms—a Synthesis problem. Nevertheless, the parts of the system are beginning to emerge. As 3-D designs from nanodevices to automobiles to skyscrapers created by engineers and industrial designers become available as a library of zoomable objects, they will start forming parts of the interactive e-book library of the kind we've been discussing in this chapter.

Another interesting development that points to the usefulness and reality of zoomable e-books is the story of two California teenagers, brothers Cary and Michael Huang. Inspired by a science lesson where his seventh grade teacher compared sizes of various biological cells, Cary Huang created a scalable, interactive map of objects with a much greater range of sizes, from infinitesimal quantum foam (10^{-35} m) to the immense observable universe (10^{27} m). The map [75], which his brother Michael helped him put on the Internet, allows the user to play with the size dimension and compare

* Indeed, even biological systems are becoming amenable to synthesis from basic materials.

various objects. For example, as we zoom in and out of the map, we can see that the thickness of a human hair is similar in size to the smallest thing visible to the naked eye, or that the Eiffel Tower is as wide as the *Titanic* and almost as tall as the Hoover Dam.

By clicking on the objects, we can get short bits of information about them, learning that the water molecule is 280 picometers wide and looks like a Mickey Mouse head. We could go on with more examples, but (paraphrasing a popular saying) describing this fun, interactive app with words is like dancing about architecture. The reader is encouraged to visit the Huang brothers' website at http://htwins.net/scale2/ or watch a demo on YouTube.

The appeal of the app lies in its zoomable nature, which allows the user to explore easily a seemingly infinite array of things. Despite the lack of advanced features we discussed earlier, such as 3-D graphics and the ability for the user to discover how objects interact with each other, the app shows the great potential of new user interfaces for interactive learning.

According to *The Economist*, a similar development is taking place at Lawrence Livermore National Laboratory, where researchers are "developing software to drill into scientific animations of particle behaviour in nuclear reactions. Called VisIt, its zooming range is equivalent to zipping from a view of the Milky Way to a grain of sand" [76].

At the time of writing this book, the e-book industry is still in the beginning of its S curve. Most of the tools available for producing an e-book imply the self-publishing business model. This is very similar to the early days of the newspaper, when a creative and enterprising person like Benjamin Franklin could combine the roles of the printer, publisher, editor, reporter, principal columnist, and the salesperson. To enliven his *Universal Instructor in All Arts and Sciences and Pennsylvania Gazette*, Franklin even wrote "reader" letters to himself. That worked reasonably well in the early eighteenth century, but by the time the newspaper business reached its Efficiency phase, highly specialized professionals filled those roles.

Furthermore, many managers had to be hired to coordinate and supervise the professionals' work. A comparable transition occurred on the web, from a lone scientist coding his HTML page to a team of graphic designers, editors, copywriters, programmers, statisticians, and PR agents, among others. As the e-book business matures, it will transition into multifaceted professional domains. Today, based on our knowledge of system evolution phases, we can boldly predict that by the time it reaches the Efficiency phase, the system's Control functionality that orchestrates creation, composition, transfer, and interactive presentation of the content, including social interaction, will be essential to the system's success.

In the next chapter we will consider some typical problems and related opportunities associated with instances of Control. Although the problems may arise in the early phases of system evolution, solving them during the Efficiency stage makes the greatest impact on the system's scalability.

24 Payload Overload

Managing the Ever-Increasing Flows of Information

The information overload many of us are suffering today is an instance of the Packaged Payload overload, a typical system problem when a particular instance of the Tool receives more Packaged Payloads that it can handle.

The problem becomes a frequent occurrence during the Distribution Buildup and the Efficiency phases when new Sources, Packaged Payloads, and Distribution routes are added to the system. Instances of the Tool designed to handle a trickle of payloads are now being forced, metaphorically speaking, to drink from a fire hose. Unless Control catches up with the new capabilities, the overall system performance may not improve despite the added number and capacity of its elements.

THE ELECTRONIC PROGRAM GUIDE

Let us consider a dominant user interface design that emerged with the development of the television. In the early 1980s during the expansion of cable and satellite television, when many new programming options became available to viewers, choosing a program and recording it on a VCR turned into a big problem. Here's how it is characterized in US Patent 4,706,121, "TV Schedule System and Process" [77]:

> As the number of television stations in a metropolitan area or on a cable network has increased, a larger number of programs of potential interest to a viewer is presented. With the use of dish antennas capable of receiving direct satellite signals, the multitude of programs available to the viewer is further increased. …
>
> In the San Francisco metropolitan area, for example, there are presently 15 different cable channels that are listed by name, not channel number. A viewer will often not remember the channel number on which a given cable service is furnished, especially if that service is only watched occasionally. …
>
> Significant problems are encountered by users of VCRs as presently operated. Programming a VCR for unattended operation requires considerable skill and care. It is necessary to select the station, the day of the week, the time,

> including a.m. or p.m., and the length of the program for each program to be recorded. The process is even more complex if the user wishes to set the VCR for automatic recording of a program in the future at a given time while watching another program at the same time.

As we discussed earlier in the book, the more elements (or choices) become available in the system, the more important the Control functionality becomes. The fifteen TV channels cited in the patent looks like child's play today when users routinely have access to hundreds of broadcast and pay-per-view channels. Nevertheless, the system-level principles behind the solutions to the channel overload problem apply today. One of the most important inventive steps at the time was the transition from the printed *TV Guide* to the electronic program guide (EPG). Rather than, or in addition to, supplying TV programming information to end users, the new system sent encoded descriptions of TV programs, or their *aboutness*, to TVs and VCRs.

Let us note this pattern. As the high-level system matures along its S curve, the Control subsystem enters its own Synthesis phase. That is, new elements are being created and a new S curve starts developing underneath the larger S curve. This understanding of the recursive nature of the system model helps us identify invention and innovation opportunities even when many people think that everything has been invented already.

The EPG many people still use today is a mature implementation of the original idea to distribute content-related information through a new Control subsystem. Within the old Control subsystem the Packaged Payload was implemented as a printed text. The user had to refer to it when trying to find a relevant program on TV or programming the VCR. In the new Control subsystem, the Packaged Payload was implemented as machine-readable data distributed along with the TV signal. The data was stored in the memory of the TV or the VCR and could be used for controlling the devices: switching channels, recording programs, and so on. The improved user ability to find and interact with TV content enabled a further growth in content offered through the system. Over time, the number of TV channels grew from dozens to hundreds, thus creating another channel overload problem, which TiVo attempted to address in the late 1990s.

In the earlier EPG example, the high-level TV content distribution system and the new Control subsystem share the same physical cable (or airwaves) for implementing their Distribution functionality. We may say that the physical cable got split into multiple virtual Distribution routes, most of which were dedicated to the content, but at least one served in the Control subsystem. As the Control subsystem matured, it acquired its own physical Distribution implementations. For example, in the original TiVo system, EPG data was distributed to the recording device through a dedicated dial-up connection. According to some of our inventions (US patents 6,611,654 and 7,403,693 [78, 79]), the EPG data was distributed to any device within the Control subsystem over the Internet. The insight behind the inventions was that mature systems (in that case it was the Control subsystem) typically have their own, rather than shared implementations of the Distribution functionality. Moreover, Control elements show the ability to bridge and sometimes go beyond their original higher-level systems. In the next subsection we will consider an industrywide transformation started as Synthesis of the Control subsystem.

DHL EXPRESS: THE SIGNAL BEFORE THE PAYLOAD

The success of Amazon, eBay, and other Internet commerce services would be impossible today without the existence of efficient express shipping services. For a small fee we can have a late birthday present delivered overnight practically anywhere in the continental United States. At first glance, express shipping has nothing to do with the Control functionality, which is the main subject of this chapter. But if we looked back at the recent history, we would realize that today's global delivery business model was originally developed as an instance of the Control functionality in overseas-bound logistics operations.

According to the *Encyclopedia of Entrepreneurs* published by *Entrepreneur* magazine [80]:

> In 1969, Adrian Dalsey, Larry Hillblom and Robert Lynn (D, H, and L) founded DHL as a service shuttling bills of lading between San Francisco and Honolulu. Hillblom had used a college loan for his share of the initial investment into DHL. All three partners, as well as friends, carried their cargo in personal suitcases between destinations using airline tickets purchased with their credit cards. [p. 147]

Why did somebody need the services of DHL to send bills of lading from Hawaii to California? What was the problem they were trying to solve?

In the 1960s, shipping from the United States to destinations in Southeast Asia increased dramatically because of the Vietnam War. An earlier invention of a specialized container ship (the Package Payload) and standard secure steel container (the Packaged Payload) that could be easily moved between ships and trains created a revolution in the shipping industry. A large number of containers could be stacked up on a ship, transported to a port, and unloaded directly onto a freight train or a truck [81].

Malcolm McLean, formerly a trucking company operator, started the transportation revolution in the 1950s.

> Malcom McLean had no maritime background, and he was in no sense an enthusiast about ships and the sea. He would often refer to his container ships as "nothing but wheelbarrows," and it may well have been his lack of any appreciation of the heritage of steam and sail that allowed him to decide calmly that he wanted a fleet of container ships that could travel the seven seas at thirty-three knots. [82, p. 117]

The new way of moving cargo was much more efficient than traditional bulk and small-lot shipping. "[S]tays in port that once were measured in days and weeks were reduced to hours" [83]. To take advantage of the opportunity, sea shipping and railroad companies built ships and cars to accommodate container-based operations. These developments brought about a fundamental change in international trade. As the volume and variety of goods shipped exploded, so did the amount of customs work that had to be done at international ports.

Before the rapid growth in shipments, when it took days to unload the ship, it was acceptable to spend that much time to process the paperwork. In the new container-based system, where load–unload operations shrunk to hours instead of days, document processing became a bottleneck. Ships that brought their bills of lading along with the cargo had to wait in or near the port while the officials reviewed their legal papers. With San Francisco and Honolulu being major destinations in the shipping triangle between the US West Coast, Vietnam, and Japan (or Hong Kong), it made a lot of sense to fly in the customs paperwork for processing days before the ship arrived. The DHL entrepreneurs recognized the opportunity and stepped into the gap created by the revolution in shipping.

A BRIEF SYSTEM ANALYSIS

Let's take a step back and describe the developments using the S curve and the system model terms. The transition to container shipping was the Distribution Buildup stage because it significantly increased the volume and variety of Packaged Payloads delivered between different destinations. But the method of delivering and processing information about the new payloads did not change to match the increase in system performance. As the system grew in scale, the Control became a bottleneck in the transition to the Efficiency stage. Therefore, a new Control subsystem was needed to sustain innovation in the higher layer Distribution. The DHL entrepreneurs created a new Control implementation by grafting it onto the existing airline transportation system. Over time, as international trade expanded, the Control subsystem grew and developed a life of its own, branching into express delivery and communications.

25 Anticipating Control Problems

The Telegram before the Train

The Control-related innovation opportunities described previously are not unique. In the nineteenth century, the rapid growth of railroads created opportunities for telegraph technology. Because every town had its own timekeeping methods, there emerged a need for clock synchronization between different train stations, so that timetables worked across the entire railway network. Also, steam engines and railroads at the time were prone to breakdowns. Despite attempts by railroad companies to maintain coherent train schedules, nobody could figure out where the train was when multiple accidents disrupted the normal course of events. Since neither railroad personnel nor passengers had any information about the disruptions, waiting for a train to arrive according to a preset schedule was a frustrating experience. As an illustration, here's an excerpt from an 1853 letter to the editor of the *New York Times* describing a common railroad problem and a possible solution:

> It would seem, from almost daily reports that our railroads are fast becoming instruments of human slaughter, and to a terrific extent. Surely, accidents to moving trains upon railroads are supposable.
>
> What is the use of a time-table unless its observance is required absolutely and unconditionally—no officer of the company, even superintendent, conductor, or engineer, being allowed any discretion in its observance, unless upon positive information about the state of the moving train.
>
> How is this information to be obtained readily? Simply by the company owning and controlling a telegraph line the full length of their line of railroad with a telegraph station at every station or stopping place of the trains—by which the exact position and condition of all moving trains can be understood at any given instant ... therefore, casualties, detentions, or any other contingency, can be communicated instantly to all points, when such information is necessary for the safety of the moving trains. [84]

Profitability of railroad companies critically depended on access to accurate information about train movements and delays. As a result, telegraph-related improvements followed closely the growth of the railroad business. Similar to the previous DHL example, growth in the Distribution led to the development of a Control-enabling information technology. To improve system efficiency by coordinating interaction between different elements, the Control subsystem needs information about them, for example, tracking Packaged Payloads prior to their arrival to the Tool.

In honor of the unknown *New York Times* reader, we call the system pattern where information about the payload is obtained before the payload reaches its destination *telegram before train* (TBT). To all the examples mentioned previously we can add the original use of radio in sea shipping and radar in aviation. One of the latest manifestations of the TBT pattern is email or short message service (SMS) notification of the recipient prior to the delivery of the order. Similarly, many websites enable tracking commercial airplane arrivals and departures so that you don't have to waste your time at the airport if the plane is delayed. Passengers checking in online for a flight or a train ride help transportation companies make their operations more efficient and reduce processing costs.

To streamline large-scale logistics systems, operators use microchips with embedded tracking data that can be written and read remotely. For example, since the beginning of the 2000s, Walmart has started requiring all its suppliers to use radio frequency identification devices (RFIDs) for tracking shipped pallets and even individual items.

In information systems, the *telegram* is often called *metadata*. Within such systems, it becomes an instance of the Packaged Payload. For example, in an email the text plays the role of the train, while its attributes, including source and destination addresses, or generated summary, play the role of the telegram. When we mark a message as spam, it helps your email system automatically produce the spam attribute of the *telegram* before the next spam *train* arrives. Based on the telegram, the email system dumps the spam directly into the trash folder.

In a social networking environment, even more sophisticated content delivery options are possible. Amazon encourages users to attach their ratings and reviews to a product page. Facebook's Like button helps friends telegram each other about a movie or a web page. In programming, engineers include high-level directives to help computers compile, set up, and execute their code. In media and politics, advanced notices are issued to alert the public about an upcoming event. In general, metadata telegrams (or *aboutness* in system terms) are widely used for enabling scalable, efficient processing of information.

In the next section we will discuss a certain class of problems frequently arising when the system needs to generate telegrams as an element of the TBT pattern. The increase in such problems potentially signals an impending step in system development.

26 System Efficiency
Solving Detection Problems to Improve Control

During the Efficiency stage, the Control becomes more sophisticated by detecting various states of system elements (e.g., properties of the Packaged Payload). Furthermore, the Control stores and analyses this information to anticipate and optimize the system's performance.

> "They have finished Phase One: reconnaissance," Lio concluded, "and yesterday began Phase Two: which is—who knows?"
>
> "Actually doing something," Barb said.
>
> **—Neal Stephenson, *Anathem* [85, p. 311]**

Consider, for example, an electronic program guide (EPG) implementation by the TiVo digital video recorder (DVR). The device enables its user to mark a TV program she likes or dislikes using the Thumbs Up and Thumbs Down interface (Figure 26.1). Once the user's personal ratings are known, the DVR begins offering automatic scheduling. The more thumbs up you give to a TV program, the higher it is placed on your TV recording list. TiVo's intuitive approach to detecting user preferences helped solve the problem of complicated VCR programming. Once the user allows the system to detect her preferences, she is guaranteed to have her favorite programs recorded. In system terms, by solving the *detection* problem, the Control sets up the recording Tool, so that its performance is optimized for the user's needs.

Facebook's *Like*, Google's *+1*, and Amazon's star buttons all serve a similar purpose of detecting user preferences. Once the system acquires the information, the Control combines it with more user-related metadata and optimizes the flow of content, including ads, news, and product notifications. The more the system knows about user interactions and the more sophisticated its data analysis algorithms are, the more control it has over what the user is going to

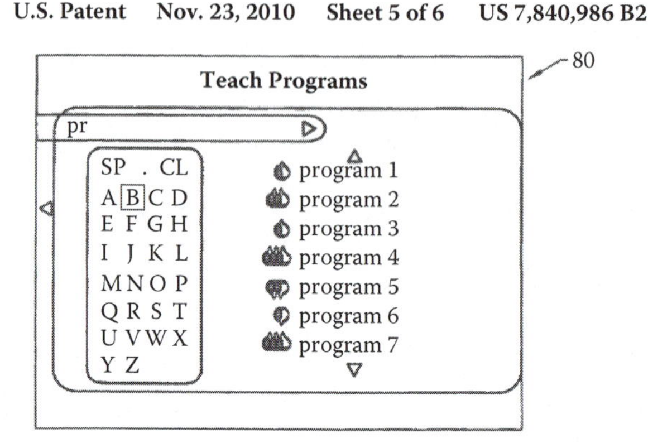

FIGURE 26.1 TiVo Thumbs Up and Down interface, US Patent 7,840,986 [86].

see in the future. Note, that content providers really want their pages to be *like*-able. That is, for the content to be consumed in volume, it has to be detectable and shareable through various social networks.

Moreover, to optimize ads and other commercial offerings, companies track user content preferences across multiple websites. Remarkably, while a plethora of tools exists for collecting personal data for the benefit of commercial entities, few options are available to the user to control the use information about her own preferences. From a system perspective, we can see strong Control asymmetries being created in large-scale web and app implementations. These asymmetries allow information providers to set up and shape the world in certain ways, while leaving to the user a limited range of control options within such a world.

In contrast with social networking, where most of the people want to be detected, the military chiefly operates in situations where everybody prefers to hide their activities. For example, steganography (from the Greek *steganos*, "covered," and "graphei," writing) is the art of concealing not only the

> Pregnant women and new parents, after all, are the holy grail of retail. There is almost no more profitable, product-hungry, price-insensitive group in existence. It's not just diapers and wipes. People with infants are so tired that they'll buy everything they need—juice and toilet paper, socks and magazines—wherever they purchase their bottles and formula.
>
> **—Charles Duhigg, *The Power of Habit: Why We Do What We Do in Life and Business* [87]**

contents of a message, but also the existence of the message itself. The practice has deep historical roots. For example, from Herodotus (484–425 BC) we can learn about covert information delivery operations performed in ancient Persia:

> One Histaieus wished to inform his friends that it was time to begin a revolt against the Medes and the Persians. He shaved the head of one of his trusted slaves, tattooed the message on the head, waited till his hair grew back, and sent him along. It worked: the message got to his correspondents in Persia, the revolt succeeded. [88]

Not surprisingly, detecting enemies and decoding their messages has been a major military intelligence task since the invention of warfare itself.

A British documentary about a successful World War II project to break German ciphers claims, "Intelligence wins wars. The more information you have on your enemy, the more you can anticipate his moves" [89]. Alan Turing, a brilliant mathematician and the inventor of the foundational principles of modern computer science, was instrumental in designing algorithms for deciphering intercepted enemy messages encoded with Enigma, a German coding machine considered unbeatable at the time. The breaking of the Enigma code allowed the United Kingdom and United States (among other military successes) to intercept and destroy German submarines that were disrupting World War II shipping routes between the two countries. The intelligence information obtained through decoded radio intercepts saved not only the lives of thousands of sailors, but also the lives of soldiers and civilians in Europe who depended upon American food and armor supplies.

Unlike the ancient Persians, the cryptanalysts in World War II Britain had access to a lot of enemy messages transmitted over the air by different German army and navy units. The researchers tested their deciphering techniques against this large body of information until they could find *cribs*, portions of text that could have a common, repeating meaning, like "one" or "weather." Using the cribs, they managed to reverse engineer German encoding techniques and decode the rest of the text. When we encounter a detection problem with an instance of the Packaged Payload we can't modify, we should think about possible cribs, naturally occurring attributes that can help us make the right identification. In addition to that, correlating cribs across a large body of payloads helps improve our detection solutions.

When the British cryptanalysts could not find naturally occurring cribs, they invented a method called *gardening*. They would have the Royal Air Force (RAF) conspicuously drop mines in specific sea areas, inducing the enemy to produce radio messages that are likely to contain location information, thus supplying the cribs necessary for figuring out the codes. As we can see, solving a problem with no ability to modify a payload directly and having no known detectable attributes may require modification of the environment or the source from which the payload originates. The goal is to get the payload to reveal itself in a reliable way.

Today, many government and private organizations mine electronic communications for valuable information. For example, according to "Danger Room," a *Wired* magazine blog covering security issues, both CIA and Google Ventures invested in Recorded Future, a Massachusetts company specializing in real-time information analysis—open source intelligence (OSINT) [90]. One purpose of the analysis would be to detect emerging conflicts by monitoring websites, RSS feeds, and messaging systems. Another goal was to determine effectiveness of an advertisement campaign, an approach very similar to the gardening method developed by British cryptanalysts decades ago. The more information that flows through the communications channels, the greater the need to find the right piece of information at the right time. When this piece is found (we solve a detection problem), it needs to be delivered to the right decision maker, either human or computer [91]. Simply put, detection problems are usually precursors to an emerging Control logic.*

As the system matures, it begins to operate within a set of structured scenarios (routines or habits), which are based on the analysis of real-time and recorded aboutness. By creating and testing new actions, we further improve the system's performance. Eventually, various feedback loops automate the system. It becomes increasingly predictable and enters the Deep Integration stage of the S curve.

* Another example of a detection/control problem is related to the growing obesity epidemic. With a great abundance of food (in particular, sugar-laden and carbohydrate-rich food "enriched" with salt and fat), making the "right" dietary and purchasing decisions becomes an impossible task for most people. Having abundant information (nutritional facts printed on the package) does not solve the problem, because this information is essentially devoid of simple, effective, decision-enabling logic guiding the behavior of the consumer.

27 Stages of System Evolution
Deep Integration

The Deep Integration stage at the top, flattening part of the S curve is characterized by rigid internal constraints imposed by a limited set of cost optimizing scenarios. Since the types of elements in the system and the ways they interact are well understood, there is no reason to invest in exploration and new application development. At this stage, the majority of instances of system elements turn into commodities. The system becomes focused on squeezing out the costs by deploying lean manufacturing methods [92], such as Six Sigma [93], and reducing innovation-related risks.

For example, by the end of the 1990s, in the analog terrestrial television system the carrying capacity of each TV channel was fixed, video formats were fixed, and end-user devices, such as TVs, VCRs, and cameras, were feature-complete and cost-optimized for value pricing in retail. Manufacturers produced devices in large volumes and at the lowest possible cost. The mode of interaction between all the elements was reduced to the "couch potato" scenario. Even when digital content was available, for example, from a personal computer or a digital video player, it had to be converted back to old analog formats, often with a loss of quality and interactive capabilities.

At this stage, one of the typical innovation mistakes is attempting to force a change in system behavior by introducing a new functionality. Let us consider the failure of the TiVo digital video recorder (DVR) to become a mass-market product in the late 1990s and the early 2000s. The device had a number of very attractive end-user features: video time shifting, intuitive program recording with advanced EPG, updates over the Internet, and others. It allowed users to skip commercials, record programs without worrying about channel numbers and clock setup, find programs with favorite actors, and much more. Early users loved the device, but it failed in the broader consumer market. Why?

Unlike TVs and VCRs of that time, TiVo was a digital computer with a lot of sophisticated software and hardware packed into a small box. To perform its amazing video feats, the device had to take as an input analog video signals, convert and compress the signal into a digital format, store the converted digital content on an internal hard drive, and convert the content back into an analog format when the user wanted to see the program. It's a lot of computing work just to time-shift the standard analog video.

Furthermore, to get EPG data, the device had to be connected to the Internet, which at the time was mostly done through a phone line. Periodically, it had to dial-up its server and download up-to-date programming information. The information was supplied in digital form, but to be compatible with the user equipment, it had to be converted into analog on-screen images. Moreover, all user interaction with the recorder, TV, set-top box, and VCR had to be performed using the analog remote control. If one of the devices was not set up right, or received a wrong control signal, the entire system wouldn't work. Besides, if the user did not have a phone jack near the TV, she had to drag a 30-foot phone wire across the house or apartment, an exercise that would look utterly ridiculous in today's age of wireless networking.

The TiVo DVR was a marvelous piece of video technology, but it did not fit into the mature analog TV system. The digital device had interface mismatches with all system elements, the key ones being the Packaged Payload (digital vs. analog) and the Control (one-way remote vs. two-way network control). The original idea to give users control over what they could do with their TV content was great. Initially, the device attracted a significant number of early adopters willing to pay for the service and put up with all the technology problems caused by the system-level mismatches. Nevertheless, the system TiVo tried to reinvent was rigidly optimized for a totally different user experience. This radical technology innovation had little hope of winning right from the start.

Ten years later, when digital television (HDTV) and wireless Internet access became the norm, TiVo-like functionality started appearing inside many set-top boxes. There was no need to perform multiple format conversions, drag wires across the room, or study user manuals to figure out how to synchronize the recorder with the tuner. Moreover, synchronization between multiple digital recorders located in different rooms turned into an automatic procedure because devices could easily talk to each other over the local (or the cable provider's) network. The TiVo DVR was an excellent idea, but it was too early for its time (or too late for a particular system) because it was applied to a system at the wrong stage of the S curve. Mature systems in their Deep Integration phase are highly resistant to innovation, and even the most creative teams that use advanced invention techniques have a hard time making a difference there.

In early 1999, ABC *Nightline* approached IDEO, a highly successful innovation firm in Palo Alto, California, with a concept of a TV show designed to open up the process of real-world innovation to millions of viewers. To make the program more exciting, ABC came up with the toughest technological challenge it could imagine: In just five days, redesign something as old and familiar as the shopping cart invented by Sylvan Goldman sixty years earlier (Figures 15.1 and 15.2). IDEO agreed, and on Monday, July 15, 1999, TV cameras started rolling. IDEO engineers and designers began their *deep dive*, a series of all-out, intense problem-solving sessions. The sessions included real-time observation of retail shoppers with their shopping carts, brainstorming for new ideas, team problem solving, and designing prototypes. To provide an insight into IDEO's creative process, ABC TV crews painstakingly recorded all these activities.

The IDEO team performed well beyond ABC's expectations. By Friday, a totally redesigned shopping cart was ready for the final show. The old big basket on wheels concept was gone. The new IDEO cart got a futuristic-looking metal frame, with six smaller hand baskets nested on it. The cart also had a child seat with a play tray and a snap-on safety bar, a scanner for direct payments, cupholders, and several other high-tech features. It was clear to everybody, both participants and observers, that the mission to show the world how to redesign an old and familiar object had been accomplished.

The ABC's *Nightline Deep Dive* TV program featuring IDEO's work became a huge media success. Ten million people watched it on the first broadcast night and millions more turned to it when it was rebroadcast a few months later. Within a year, the novel shopping cart won a Silver award from the Industrial Design Society of America. The term *deep dive* became a common name for a process of intense creative work for developing new ideas. After getting all this unprecedented publicity from a major US TV network and receiving a prestigious award from a highly respectable professional society, the new shopping cart seemed to be destined for great commercial success.

But the innovation did not materialize. Six years after the *Deep Dive* show aired on ABC, a blogger posted a question, "What happened to the IDEO shopping cart?" Nobody knew. Except for a few YouTube videos and corporate web pages, the cart had disappeared from the public's view. Despite all the past excitement about the creative process, high-end supermarkets and grocery stores did not rush to order the new shopping cart. A great idea had not become a successful innovation.

Why was that? How come a modern design marvel produced by the best design talent had lost its market battle to a 1937 invention of Sylvan Goldman we discussed earlier in the book? Was 1999 a particularly bad year for shopping cart innovation, or was there a broader pattern of innovation failure?

To answer these questions, consider another shopping cart invention patented in the same year [94]. In this case, the idea was developed by Amazon.com, an online retailer. In sharp contrast with IDEO's hardware solution,

Amazon.com created, implemented, and patented an improved virtual shopping cart (Figure 27.1). Their problem-solving process was not televised nationally, but its outcome produced enormous value for the company in the emerging e-commerce industry. Moreover, the virtual shopping cart with the one-click purchase option became a dominant design in the world of e-commerce. In both cases—IDEO and Amazon.com—a strong creative team invented a version of the shopping cart. But the difference in innovation outcome, an industrywide adoption versus a design award, was due to the stage of the S curve where the teams applied their creative efforts.

That is, the Amazon.com team wanted to improve an e-commerce system that was entering the Early Growth stage, where every system element provided plenty of opportunities for improvement. Moreover, enabling users to buy more online was a perfect fit with the *more is better* guiding principle typical for that system evolution phase. In contrast, ABC asked IDEO to redesign a very mature shopping cart and implicitly, the entire brick-and-mortar shopping experience. Namely, how consumers pick groceries at the supermarket, how they pay for the purchases, how they take their purchases to the parking lot, how the supermarket employees collect the shopping carts from the parking lot, and so on. Although the redesign created good entertainment value, it failed to target cost reduction, the essential problem for a system in its advanced stages of the S curve. While some inventions can be too early for their time, the physical shopping cart redesign was too late. The choice of the problem, not the creativity of the team, determined the bad innovation "luck" of IDEO's shopping cart.

U.S. Patent Sep. 28, 1999 Sheet 6 of 11 5,960,411

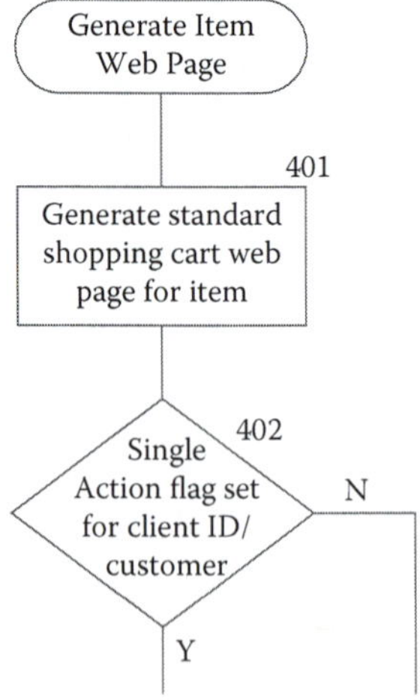

FIGURE 27.1 Example: Reinvention of the brick-and-mortar shopping cart.

28 Choosing the Right Problem

Matching Innovation Targets with Stages of System Evolution

One of the fundamental attributes of innovation-related problems is choice, almost always surrounded by the uncertainty of the outcome. Edison did not have to take on the task of bringing electric lighting to Manhattan. Instead, he chose it as a technical and business challenge. George Westinghouse didn't have to go into the electric power business and challenge Edison's creative genius. Similarly, Henry Ford did not have to build a new type of car and a new mass manufacturing system; it was his own free choice. Jeff Bezos did not have to develop Amazon.com, an online book-selling service, and move to Seattle from New York, where he was having a successful career developing electronic banking technologies. It was a matter of choice, not routine.

Choosing the right challenge has long been considered as an essential skill for inventors and entrepreneurs. Dr. R. Keith Sawyer, the author of *Explaining Creativity*, notes this pattern when he writes, "[T]he most important characteristic of creative people is an almost aesthetic ability to recognize a good problem in their domain."

When startups approach venture capitalists (VCs) from Kleiner, Perkins, Caufield, and Byers, arguably the most successful VC firm of all time, the first question they receive is about their team, its background, and its members' skill set. But the second question is most likely to be about the problem they chose to solve. Along with the specific technical or business skills, the quality of the problem determines the startup's chances for raising funds for implementing their idea and increases their probability of creating a successful innovation.

Successful entrepreneurs and innovation managers are acutely aware of this issue. For example, in their talk at Stanford University's Entrepreneurship program, Kevin Systrom and Mike Krieger, the founders of Instagram, a popular mobile photo sharing startup, said that finding the right problem to solve was the hardest part of their technology and business effort [95]. In another example, Judy L. Estrin, the author of *Closing the Innovation Gap*, describes a product failure study in the early days of Intuit, a leading accounting and business software firm:

> *[W]*hen Intuit introduced eight new products that flopped between 1994 and 1999, founder Scott Cook was determined to find out why. Analyzing these failures, he was surprised to discover that the same employees, types of technology, brand names, distribution channels, and target customers were often involved in both the company's flops and successes. The difference was that "the successful products addressed an important unsolved customer problem," he says. "Nobody has tried to solve the problem because they didn't see it." [96]

Often, customers themselves don't see the high-value problem until they get a chance to experience the innovation firsthand. John Scully, a former CEO of Apple, described one of Steve Job's most important business principles: "He [Jobs] said, "How can I possibly ask somebody what a graphics-based computer ought to be when they have no idea what a graphics-based computer is? No one has ever seen one before." He believed that showing someone a calculator, for example, would not give them any indication as to where the computer was going to go because it was just too big a leap." Similarly, when asked what market research went into the iPad, Mr. Jobs replied: "None. It's not the consumers' job to know what they want" [97].

Economist and best-selling author Clayton Christensen notes in his seminal book *The Innovator's Dilemma*, "the maxim that good managers should keep close to their customers can sometimes be a fatal mistake" [98, p. 4]. He shows that in high-tech innovation, highly successful companies whose products follow their customers' needs often fall victim to disruptive technological changes.

Meanwhile, successful entrepreneur and educator Steve Blank, who teaches entrepreneurship at the University of California–Berkeley and Stanford University, advocates close involvement with customers through what he calls the *customer development process*. In his book *The Four Steps to Epiphany*—required reading for university startups seeking funding from the National Science Foundation—he states his core principle, "[T]he 21st century products developed with [startup] founders who get out in front of customers early and often will win" [99].

Although Steve Blank targets his advice toward creators of virtual, software-based products, Malcolm Gladwell [100], another best-selling author who has written extensively on innovation, shows that inventing a new spaghetti sauce or soft drink should be grounded in insights gained from a deep understanding of the customers. He tells us the remarkable story of Howard Moskowitz, a market-research scientist and direct descendant of an eighteenth-century Hasidic rabbi, who was instrumental in creating a customer choice revolution in the food industry.

Faced with the task of discovering a formula for a perfectly sweetened Diet Pepsi, Moskowitz performed a typical consumer choice survey, creating sample drinks with every conceivable variation in sweetness and testing them on hundreds of people. The resulting data was all over the map with no clear indication of a single preferred customer solution. The failure to discover a "silver bullet" led Moskowitz to an important insight. Instead of a perfect Diet Pepsi, there should be multiple drinks with various levels of sweetness. Although his original corporate clients rejected the idea, Moskowitz persisted. After several years of explaining his concept to the food industry, he caught a break when the Campbell Soup Company, desperate for a new successful product, asked him to do customer research on a new spaghetti sauce.

To cut a long story short, Moskowitz created 44 varieties of the sauce and tested it on hundreds of people in different US cities. He discovered that customer preferences fell into different categories, one of which was left untapped by the market. In 1989–1990, based on Moskowitz's research, Campbell launched extra-chunky Prego spaghetti sauce, which turned out to be extraordinarily successful. Other successful consumer research projects followed, and eventually, customer segmentation became the new golden rule in food marketing.

Today, we can find variety everywhere, from thirty-six kinds of spaghetti sauce in the supermarket to dozens of combinations of coffee cup sizes and syrup flavors at Starbucks. Attending to customer needs, be it with data bits or tomato chunks, seems like a winning formula for innovation, which seems to directly contradict the Steve Jobs' approach described earlier in the chapter.

With so many convincing recommendations on a successful strategy, the situation is reminiscent of an old joke:

> In the small villages of Eastern Europe, the rabbi was the undisputed authority among the Jewish people. Once, two men who had a major dispute decided to seek a resolution from a village rabbi. The older man went to the rabbi first and carefully outlined his side of the argument. The rabbi listened intently and finally said, "You are right." The man went away satisfied. Later in the day, the other party to the dispute arrived and told the rabbi his side of the story. The rabbi again listened carefully, was impressed with the arguments, and replied after some thought, "You are right." Later, the rabbi's wife, who had overheard

the rabbi's conversations with both men, said to him, "Rabbi, you told both men that they were right. How can this be?" To which the rabbi replied, "And you are right too!"

From a system perspective, market research (asking consumers about their preferences) represents an attempt to solve a *detection* problem. That is, by presenting potential customers with a sample product or a service we are trying to detect its desirable and undesirable attributes. Having solved the detection problem, we can then try to solve a Control problem. That is, we want to produce and deliver a product or service with desirable attributes to a relevant group of consumers. But if the system as a whole doesn't work, proposing a solution to the detection problem would be premature. That is, if we can't produce or deliver the goods in a cost-effective, reliable manner, even the most perfect implementation of the Control subsystem won't produce the desired outcome.

For example, as Malcolm Gladwell tells us, the key assumption behind Howard Moskowitz's extra chunky sauce idea was that his client, the Campbell Soup Company, had a system in place that could easily produce and deliver the sauce to spaghetti-loving consumers. It may sound trivial, but for the sauce to become "lucky," the following pieces of the food retail puzzle had to exist: factories capable of producing and packaging a variety of sauces; distribution channels for delivering the packaged sauces to supermarkets and restaurants; supermarkets and restaurants with sufficient shelf space for a broad variety of spaghetti sauces; lots of consumers who liked to eat spaghetti; bottles, containers, or other food preservation technologies capable of maintaining the desirable qualities of the sauces; an effective advertisement campaign to attract consumer attention to the new sauces; pricing and consumer demand business models; money that consumers pay when they buy the sauce, and so on.

Because the system was largely in place, adding another kind of spaghetti sauce was a relatively easy task. Furthermore, Moskowitz's recommendation for the new product had a clearly identified *aboutness* attribute, chunkiness. By comparison, the earlier consumer study to find preferred sweetness for the Diet Pepsi did not result in a specific product recommendation, which made it difficult for the company to implement the idea. The difference between the two cases was the ease with which producers, retailers, and consumers could solve their own *detection* and *control* problems. Putting one extra kind of sauce on the shelf was simple. Presenting consumers with forty-four different choices right away would create confusion, a situation observed in the early smartphone market. Later on, once the concept of consumer choice turned into the norm, adding more and more choices to the system helped produce growth in consumption. Within such a system, listening to

> A certain style of product marketing was invented here [in Silicon Valley of the 1970s]. During this time it was apparent that if we went to the marketplace and asked people what they wanted, they would give a very complete list of our competitor's parameters. But a company would never get into the passing lane with that information. All great products arose not from customer research, but from product marketing and a full understanding of what customers were really trying to do.
>
> **—E. Floyd Kvamme, "Life in Silicon Valley" [101]**

customers, identifying their preferences, and targeting new product segments was a smart innovation idea.

By contrast, a typical Steve Jobs innovation attempt involved the creation of a new product category or an entire system. In the late 1970s, Steve Jobs and Steve Wozniak developed a personal computer (PC) that targeted consumers, rather than technology enthusiasts. In the early 1980s, Apple created a category of consumer PCs with a graphical user interface (GUI). When Jobs left Apple, he tried to create NEXT, a new type of computer workstation for university researchers. NEXT failed in the market, but its software architecture and programming language eventually became the foundation of new operating systems for Macs and iOS devices. In the 1990s, another Jobs venture, Pixar Animation Studios, produced *Toy Story*, the first computer-animated feature film. In the early 2000s, Steve Jobs and his team created a system of products and services for mobile communication and entertainment devices—the iPod, iPhone, and iPad. Around the same time, Apple introduced a new type of consumer shopping experience in their stores.

Most, if not all, of these innovations targeted early stages of the S curve. Under the circumstances, asking consumers about their preferences for a specific feature of a new product would not make a major difference because the success of the system depended on its performance as a whole. Besides, Steve Jobs himself was the most demanding user of his company's products. He was famous for not only his grand and inspiring visions of the future and persuasiveness, which collectively came to be called the Reality Distortion Field, but also for great attention to implementation detail. For example, Vic Gundotra, a senior VP at Google, recalled an episode when Steve Jobs called him one Sunday morning, January 6, 2008, with an urgent issue:

> I've been looking at the Google logo on the iPhone and I'm not happy with the icon. The second O in Google doesn't have the right yellow gradient. It's just wrong and I'm going to have Greg fix it tomorrow. Is that okay with you? [102]

Note that this is the level of scrutiny Steve Jobs applied to an app provided by Google, a business partner involved in the iPhone launch. The spaghetti sauce equivalent of this would be fixing the shade of the color of the product label on a bottle sitting on a store shelf among dozens of other bottles. It is highly unlikely that consumers evaluating the taste of a new sauce would be so critical. It is even less likely that they would be able to identify key success factors behind the iPhone. These factors include the look and feel of the device, as well as invisible ones, like the cellular data network, new content services, the App Store, and others.

As we discussed in the beginning of the book, people are prone to the *what you see is all there is* (WYSIATI) effect. That is, they tend to focus narrowly on what is available to their immediate perception. The great achievement of Steve Jobs and the rest of his team at Apple was the ability to see the new system as a whole and use this vision to work out minute implementation details for its key elements. Once the new system was created, it provided a new environment for consumers to express their preferences regarding specific features of a system component. For example, using the iTunes and App Store, iPhone users were enabled to rate and review songs, apps, books, podcasts, videos, and other content items. In essence, the users became a multitude of virtual market research panels for Apple and their partners.

Although business models are outside the scope of this book, it is worth mentioning here that the system was set up in a way that allowed Apple to make money on every new "sauce" sold for the iPod, iPhone, or iPad. Because the company reportedly takes a 25–30 percent cut of revenue from every transaction, it is in their best interest to flood the system with purchase-triggering choices and encourage intense competition among app developers and content providers. With hundreds of thousands of apps and songs in the store, the users are bound to discover something they like, helping Apple generate approximately $2B in revenue from iTunes per quarter (Q2, 2012) [103].

However, for developers, discovering success is a far more difficult task. According to *ArsTechnica*, 60 percent or more of them don't break even on their development costs, 68 percent earn $5,000 or less with their most successful app, and 80 percent don't generate enough revenue to support a stand-alone business [104]. Under the circumstances, Steve Blank's advice to developers to keep their costs low and "get in front of their customers early and often" makes good business sense. Although, it hardly guarantees the success on the scale of the iPhone.

One may argue that Steve Jobs' creations are a work of genius and the Apple setup is unique. Therefore, it would be impossible to re-create Jobs' recipe for success, which puts strategic vision ahead of immediate customer feedback. But when we look at another company whose business model involves creating a shopping environment, Amazon.com, we see a similar pattern. For example, analytics measuring revenue from 11 million users show that over a sample period of 45 days in early 2012, for every dollar made in the iTunes App Store, there were 89 cents made in Amazon's Appstore selling Android apps [105]. When we consider that at the time of the survey the Amazon Appstore had been in business for less than a year, the results are even more impressive.

By contrast, Google Play, the equivalent of an Android app store, made a prorated 29 cents, despite the fact that Android is Google's own mobile operating system and Google started distributing apps much earlier than Amazon.com [105]. In other words, Android apps are much "luckier" with Amazon than with Google. The data indicates that Jeff Bezos and his Amazon.com team are talented enough to mirror Apple's business success. Specifically, their success involves a strategic move to put together a shopping system where lots of choices are available to consumers and they can reveal their shopping preferences and act on them.

Similarly, to achieve market success 100+ years earlier, both Thomas Edison and George Westinghouse had to experiment with specific electric devices, such as the electric light bulb and the AC motor. But their experimentation with consumer acceptance of their new technology solutions was set within a certain system vision. As we discussed earlier in the book, Edison made a large bet on a novel DC-based low-current power distribution system that could span several city blocks. The time, money, and effort the inventor spent looking for a high-resistance light bulb filament could result in success only if his entire system worked. Because his customers (remember the WYSIATI effect) could appreciate only the instantaneousness, brightness, and longevity of the light bulb, Edison could not rely on them to provide him with a proper vision of an electricity distribution system.

Likewise, Westinghouse could rely on his customers to express their preferences for a specific power generator or an electric motor, but first he had to imagine how to build a large-scale AC-based system capable of delivering a lot of electric power over long distances from the generators to the motors.

Because innovation outcomes are uncertain, we may consider them as experiments that can either succeed or fail. If an experiment fails, which is a typical early innovation experience, something in the system has to be changed. As many experimenters know, tweaking one parameter of one element within a certain fixed setup, like spaghetti sauce chunkiness, the resistance of the light bulb filament, or the color of a mobile app icon, can be easy. On the other hand, changing the setup itself or even changing multiple parameters for multiple system elements often creates more confusion than clarity. As with any other complex problem, the innovator's dilemma, whether to listen to one's customers or not, doesn't have the right solution. Instead, the solution can be better or worse, depending on the complexity and the costs of running a series of innovation experiments.

From a system perspective, to decide on an innovation strategy, it is useful to determine which system element we would like to target. Then we need to understand where on the S curve the element itself and the system as a whole belong. For example, if there's no system that can produce a desired outcome, or if everything in the system is rigidly optimized for a particular set of outcomes, experimenting with its elements and listening to consumer feedback may not help. As we discussed earlier in the book, a digital video recorder may have tough "luck" succeeding within an analog TV system. A redesigned shopping cart that millions of consumers would love to see at their supermarket may not become a reality in a cost-optimized retail environment.

Furthermore, Clayton Christensen has demonstrated that even when a powerful company creates an innovative solution its customers like, it might not be able to deliver the solution to the market because of internal business model conflicts. For example, a computer with the GUI was invented at Xerox PARC, but the company management was not able to capitalize on the idea. It took the efforts of Apple and then Microsoft, Intel, and IBM to create the GUI PC revolution. Again, in system terms, even when a company is successful in solving a *detection* problem with regard to customer preferences, the company may not have the right Control subsystem to take advantage of the solution.

Major innovation breakthroughs show that it takes vision beyond any consumer feedback to create a new *system*. Even then, once the system is created, it takes a lot of experimentation, technology, and business savvy to discover and satisfy consumer preferences within the system. From a system perspective, rather than following specific contradictory recommendations on consumer feedback, it is useful to evaluate configurations at different stages of their development. Also, to avoid or mitigate the consequences of the WYSIATI effect, we need to learn how to compare entire systems instead of focusing on a single difference between specific instances of a certain system element. The technique for facilitating this task is the subject of the next chapter.

29 Tech Battles
Discovering System-Level Competitive Advantages

The Tech Battles technique was inspired by the events of the so-called War of Currents between Thomas Edison and George Westinghouse. As we discussed earlier, in the late 1880s, the two innovators fought in the battlefield of public opinion which electricity technology was better for consumers, direct current (DC) or alternating current (AC). Paradoxically, for the consumers in their everyday lives there were no clear-cut, perceivable differences between DC and AC. At the time, the light bulb was the only common electrical device available to the public. The two major consumer features of the light bulb, longevity and brightness, were largely the same for DC and AC implementations.

It was not until the industrialization of tungsten in the beginning of the twentieth century that the advantages of AC light bulbs, which could last longer and shine brighter under high currents, became apparent to consumers. By that time the War of Currents was long over, won by Westinghouse on the strength of his industrial AC applications. Therefore, in the 1880s, it would be a mistake to base the choice between the two technologies on consumer preferences.* To decide where to invest his time, money, and effort, an aspiring electricity technologist or entrepreneur of the time might have been better off ignoring the public's opinion altogether.

As we discussed in another example earlier in the book, the current prevailing wisdom is that the iPod introduced to the market in the early 2000s became successful because of its superior design, developed by Apple's brilliant team led by Steve Jobs and Jonathan Ive. But if we go back to the time of the iPod launch, we discover that Apple's other products, desktop and laptop computers among them, were designed by exactly the same brilliant creative team. Nevertheless, they had less than a 5 percent share in the PC market. Somehow, the famous Apple design didn't make much difference in

* It is important here not to conflate the terms *consumer* with *customer*. The direct customers of Westinghouse were not rank-and-file consumers, but industrial clients motivated by different factors. Nevertheless, regular consumers could influence the decision process, so their (informed or uninformed) opinion had to be *managed* carefully.

that market segment despite Apple having pioneered the field of PCs with GUI. In other words, the supposed decisive role of design in determining market advantage for a consumer device doesn't tell the entire story.*

Similarly, the design of the shopping cart created by IDEO for ABC TV was superior to ugly shopping carts used in modern supermarkets. But this advantage didn't translate into an innovation success, despite overwhelmingly positive responses from millions of consumers.

In another example, when compared side-by-side with the VCR, TiVo was a superior consumer device, which nevertheless had a hard time gaining a significant market share.

Furthermore, when we go beyond devices and consider major services, Google's success in the web search business presents another paradox when we try to attribute it exclusively to consumer advantages. Today, many people believe that Google overtook Yahoo in search and advertisement services because of Google's superior search technology. Google's search results were so good, so the story goes, that users stopped using Yahoo, which dominated the web in the late 1990s and the early 2000s, and voted with their virtual feet for the new cool technology startup.

In reality, at the time of the switch, Yahoo and Google search technologies were one and the same. In 2000, Yahoo, citing Google's strong consumer focus, selected them to run Yahoo's search service [106]. In 2001, Yahoo paid Google $7.1 million for handling search queries from all Yahoo web properties. Furthermore, consumer advantages of Google technology were understood from the very beginning because its use increased Yahoo search traffic by 50 percent in two months [9]. Attributing Google success exclusively to its search engine, without considering other system-level factors, would be a mistake. Had we compared the companies as race cars, we wouldn't attribute Google's win to the performance of its engine because both "cars" used the same engine.

It seems that when we, as consumers, justify advantages of one technology over another in retrospect, we tend to exhibit hindsight bias, especially in cases when the technology results in a dominant design (see Chapter 13, Early Growth). This mistake is the consequence of a narrow mental focus directed at one particular system element, usually the Tool or the Packaged Payload, while taking the rest of the system for granted. When forecasting the success of a technology, such narrow focus results in hype and inflated expectations, because people severely underestimate the value of the system as a whole.

---

* While it is easy to ascribe the differences in the early success of the iPod versus that of the Mac to luck or design brilliance, the point is to deconstruct the situation (the system) to gain deeper insight.

This kind of thinking occurs because the majority of our life experiences come from mundane processes within mature systems. In such systems, somebody else has already put all of the system elements in place, while we are free only to select a particular commodity delivered via a well-established process. For example, when we choose between two toasters in a store, we assume that there will be a power outlet in the kitchen and there will be electricity in the outlet. When we chose between two brands of toilet paper rolls, we assume that there will be a bathroom with a toilet and a holder to hang the rolls.

However, during the early stages of innovation one cannot afford to make this assumption because products and services are still evolving; they may fail or succeed not because of their own deficiencies or advantages, but because of either a poor or good fit with the rest of the system. The paths of innovation are full of choices under uncertainty and innovators are bound to make mistakes. Nevertheless, we know that certain environments can be more punishing than others. One doesn't have to be a creative genius to realize that raising a novel breed of fish will be easier in a pond than in a crate with desert sand. Nevertheless, even the cleverest people can forget about the system working in the background. Here's how Warren Buffett, a highly successful investor, described in a letter to shareholders the difference between mistakes you can and cannot make:

> It is comforting to be in a business where some mistakes can be made and yet a quite satisfactory overall performance can be achieved. In a sense, this is the opposite case from our textile business where even very good management probably can average only modest results. One of the lessons your management has learned—and, unfortunately, sometimes re-learned—is the importance of being in businesses where tailwinds prevail rather than headwinds.
>
> **—Warren Buffett [107]**

The purpose of the Tech Battle technique is to help us discover the *winds* by comparing performance and scalability of not one, but all system elements: Tool, Source, Distribution, Packaged Payload, Control. To start with a simple example, let's replay the War of Currents between Thomas Edison and George Westinghouse (Table 29.1).

CONSIDERING THE TOOL FUNCTIONALITY

Let us begin in the Tool row. Earlier, we acknowledged that Edison's and Westinghouse's light bulbs were very similar in functionality. Because of that, the general public could not easily decide whether AC or DC technology was better. Both bulbs provided the same amount of light at similar

TABLE 29.1
Tech Battle: AC vs. DC

	System	
Element Name	**Direct Current (DC)**	**Alternating Current (AC)**
Tool	=	=
Source	–	+
Distribution	–	+
Packaged Payload	–	+
Control	–	+

costs, and the consumer could not tell the difference between them based on personal experience.

In the early days of consumer lighting, the reliability of bulbs was the biggest issue, but it depended on a particular manufacturer, rather than the type of electricity the bulb used. Once the need for a high-resistance light bulb was understood, many inventors and manufacturers created their own implementations. Edison and Westinghouse fought for years over their light bulb patent rights, but in the end it was the General Electric Corporation that became the dominant player in the light bulb market after the original patents expired.

As we discussed in previous chapters, Edison's system did have an important advantage in the Tool area. In the early years of industrial use of electricity, only DC motors could be designed and built. Westinghouse attacked this issue by working with Nikola Tesla, and eventually developed AC motor technology, which nullified Edison's initial advantage in the Tool area. To mark this system-level equality we write " = =" in the Tool row of the comparison table.

CONSIDERING THE SOURCE FUNCTIONALITY

On the Source side, the AC-based system had an advantage because AC generators were simpler in design, cheaper, and could be used in large energy production units. For an industrialist who wanted to build a big power plant, exploiting free waterpower, or a large supply of cheap coal, AC would be a natural choice.

Also, it is important to realize that this Source advantage only worked within the greater context of American industrial growth. New, more powerful AC generators were more desirable for large-scale installations, even though Edison's DC technology was more mature and energy efficient. The growth of the American economy, a higher-layer system development, provided the tailwind for George Westinghouse's AC solution. To mark this advantage, we put "+" for the AC system and "–" against the DC system. Note that to reflect the degree of advantage or disadvantage, we can use multiple plus or minus signs, or alternatively give the system certain grades, like on a scale from 0 to 5.

CONSIDERING THE DISTRIBUTION FUNCTIONALITY

The Distribution was another area where AC had a major advantage over DC. Westinghouse's system was more suitable for technology specialization and growth because it could span longer distances between energy producers and consumers. It gave American entrepreneurs more flexibility in deciding the locations of their factories, stores, offices, and houses. The technology enabled businessmen to place the Sources and Tools where the installations made the most economic sense. That is, a power plant could be built where waterpower or coal supply was cheap, while factories and houses could be constructed close to where workers lived. Furthermore, a large power plant could serve a wide variety of customers, which allowed for a degree of capital risk diversification.

From a technology perspective, high-frequency AC power lines suffered less resistance-related losses than comparable DC distribution systems. As a result, AC was a great solution for flexible delivery of energy on the new industrial scale. To mark this advantage, we put "+" for the AC system and "–" against the DC system.

CONSIDERING THE PACKAGED PAYLOAD FUNCTIONALITY

After the transformer and the AC motor were invented and implemented, a system with an AC-based Packaged Payload held a strong advantage over one with a DC-based Packaged Payload. AC-based energy was inexpensive to generate, efficient to distribute, and easy to control (with the transformer). AC is also at least on par with DC in applications, such as light bulbs, motors, and so on.

Although Edison claimed that AC was more dangerous than DC for people and animals, with proper wire insulation and other safety measures, the safety disadvantages of AC became a nonissue. Because of its great fit with the entire electricity distribution system, AC turned into a dominant electricity "packaging" design for years to come. To mark this advantage, we put "+" for the AC system and "–" against the DC system.

CONSIDERING THE CONTROL FUNCTIONALITY

In Edison's DC system, control over current flow could be done mostly statically, like with the parallel circuit design we discussed earlier. While it worked great for relatively small installations, DC power could not be easily switched up or down depending on the change in power demand. With the improvements in the transformer and switching technology, the AC advantage in the Control area proved to be very important during the later stages of the electric grid expansion. Eventually, when it was needed, large regions of the country were able to exchange electric energy, providing for a greater reliability of the energy grid as a whole. To mark this advantage, we put "+" for the AC system and "–" against the DC system.

SUMMARY OF THE TECH BATTLE TECHNIQUE

As we can see, even though Edison's DC system had an early advantage in the Tool functional area, over time its disadvantages with regard to other functional elements became apparent. Westinghouse won the War of Currents, not because he had the better technology solutions right away. Rather, in addition to appreciating his advantages in large-scale power production and distribution, he saw important holes in his own AC system—the light bulb, transformer, and motor. He proactively sought how to address them by recruiting inventors, buying patents, and building alternative solutions.

The AC-based system won the Tech Battle despite the lack of immediately perceived consumer advantages; the light bulbs didn't shine brighter and the motors didn't turn faster. Rather, its key advantages showed up in the implementations of other functional elements essential for *scalable deployment during the growth stage*. Given the strong industrial development tailwinds of that time, AC was a better technology bet.

Similarly, Edison's DC city lighting system was a better bet compared with the gas solution. It was safer, cleaner, and easier to control. In the relentless process of creative destruction, Edison's DC electricity approach destroyed gas lighting, but eventually fell victim to the next round of innovation. All of these inventors and entrepreneurs were brilliant, but the timing of their success depended heavily on the stage of system evolution in the industry in which they chose to apply their talents.

30 TiVo versus VCR
A Detailed Application Example of the Tech Battle Technique

Now that we are familiar with the technique, let us briefly consider another example—TiVo versus VCR (see Table 30.1). Since the TiVo technology is relatively recent, we will focus our attention on the timing and innovation opportunities that the Tech Battle technique helps us identify during the early days of digital video recording.

As we discussed earlier, people often make predictions about a particular technology based on the performance of one or two salient elements of the system. Moreover, marketers design technology demonstrations to emphasize the specific advantage of such an element. But the practice shows that innovation is a team competition, both literally and figuratively. A strong team, or as we called it *creative crowd*, is essential for developing and implementing breakthrough solutions. For example, when considering an investment in a startup, most venture capitalists evaluate the quality of the team first. They check not only whether individuals have relevant skills and experience, but also how well they can work together in challenging circumstances.

On the other hand, the evaluation of a technology solution is often considered sufficient if performed based on the solution's individual merits, without proper regard for its system *team*. The proper timing for the innovation is considered on the basis of calendar time (a measure of our everyday reality that reflects Moon phases or the Earth's rotation cycle around the Sun). But unless the proposed solution deals with space travel, agriculture, or other season-related issues, innovation timing has nothing to do with astronomy.* Crucially, the Tech Battle technique considers the timing issue as a function of the relationship between system elements and system constraints.

* Seasonal shopping trends notwithstanding. Moreover, in business-to-business markets, financial constraints are more prevalent. As we discussed earlier in the book, such constrains should be proactively explored using the STM operator.

TABLE 30.1
Tech Battle: TiVo vs VCR

	System Configuration			
Element Name	**Analog Video with VCR**	**Analog Video with TiVo**	**Digital Video with VCR**	**Digital Video with TiVo**
Tool	+	–	–	+
Source	+	–	–	+
Distribution	+	–	–	+
Packaged Payload	+	–	–	+
Control	–	+	–	+

The higher-level tailwinds or headwinds can work toward breaking or enforcing the constraints. Therefore, seeing the advantages and disadvantages of the system as a whole helps identify specific targets for invention and innovation efforts. Instead of predicting a precise implementation for a possible solution, the Tech Battle technique, combined with the S-curve method, focuses our efforts on predicting the right problem at the right time.

The Tech Battle we are about to show is not between the two devices, but rather between two systems that have these devices as instances of their functional elements. It is similar to a comparison between two warring parties, which involves looking beyond just the strength and potential tactics of the opposing armed forces. That is, in addition to detachments ready for an immediate battle, we evaluate their supply lines, war material production capacity, intelligence services, headquarters experience, military alliances, and so on [108].

TIVO VERSUS VCR: THE TOOL

The function of the Tool in the system was to record and play (time shift) video distributed to homes via cable, satellite, or terrestrial broadcast. Compared to the TiVo, the VCR did not need to convert the content into digital form for storage and playback. At the time of the first TiVo product introduction, VCR tapes, the storage media for analog TV, were a lot cheaper than the high-capacity hard drives necessary for storing compressed digital video. With the VCR, when the user needed more storage space, she could simply buy and insert another tape. On the other hand, to free up the space on the TiVo's hard drive, the user had to delete content. To save it, she had to hook up the TiVo to a VCR and transfer the content to the tape. Properly connecting multiple analog video devices—TiVo, VCR, TV, set-top box—was a relatively complex task. Besides, the setup meant that the user could not get rid of the VCR altogether. Moreover, compared with TiVo, the VCR was an inexpensive commodity appliance, which did not bode well for the TiVo as a direct VCR replacement.

Because TiVo was a digital computer, its competitiveness depended on the price-performance of hard drives and video encoder–decoder processors. By applying the STM operator, we can easily imagine that the best-ever TiVo would have infinite storage, instantaneous conversion from analog to digital, and zero transaction costs. A possible configuration that fits this ultimate requirement would be a server with lots of cheap storage, digital-only video, and advertisement-supported content distribution. Another implementation option would be a peer-to-peer (P2P) video distribution system. Of course, network bandwidth would become a bottleneck, but these and other ideas would be in line with the prevailing tailwinds in the information technology (IT) and consumer electronics (CE) industries.

Overall, we conclude that as a physical device (the Tool), the TiVo was not highly competitive with the VCR, despite the TiVo having features attractive to consumers. Without the hindsight bias and simply following the evolution curves of auxiliary technology, we could tell that even if TiVo succeeded initially, its lifetime in that form would be severely limited by the advent of broadband (faster Distribution) and more sophisticated server technology.

TIVO VERSUS VCR: THE SOURCE

At the time of the TiVo market introduction, most video Sources were analog. Therefore, neither TiVo nor VCR had any immediate advantages. Although satellite broadcasts were digital, satellite set-top boxes had to output analog signals because of content copyright restrictions and TV sets having only analog inputs. A set-top box with an internal TiVo functionality implementation would make a lot of sense because it could store digital content directly. Unfortunately, the TiVo as a stand-alone device could not take advantage of that particular Source right away because of the need to completely reengineer the box, and the limited size of the satellite TV market. Only when the entire content distribution and CE industries switched to high-definition television (HDTV) formats and most of the Sources turned digital, did a system with digital storage gain a clear advantage.

In summary, a video distribution system with TiVo functionality (not necessarily implemented as a stand-alone device) would have gained advantage if the industry transitioned from analog to digital. Under the initial circumstances, technology licensing would be a good long-term business strategy. Nevertheless, at the time of the TiVo introduction, the Sources didn't provide any advantages to the device.

TIVO VERSUS VCR: THE DISTRIBUTION

Within the analog video system the Distribution favored the VCR, because it was set up to deliver analog signals with embedded electronic program guide (EPG) information, which then could be used for recordings. Furthermore, if we included in our consideration movie rental services, which at the time were tape based, the VCR was a much better fit than the TiVo. In addition, a *sneaker network* with multiple VCRs, like a family or a group of friends, could easily share their favorite recordings. On the other hand, sharing TiVo recordings was either technically difficult or outright prohibited by copyright laws.

In a digital video system, the Distribution could give strong advantages to the TiVo provided the content was delivered via streaming or downloading rather than through DVDs or other physical memory devices. Also, if one could attach a digital video device with TiVo-like functionality to the network that was feeding content to individual home-scale TiVos, that network-based device could store and time-shift content for multiple local Sources and Tools [109]. That is, in a distributed system, a server-based TiVo implementation could serve many devices, both stationary and mobile. No VCR would be able to accomplish this task, even with a major redesign. With a matching implementation of the Distribution, a TiVo-enabled device had the potential to become both a content destination point and a server for other video devices. In system terms, depending on the user scenario, it could play roles of the Tool, Source, and Control.

Where the analog video system was concerned, TiVo was a misfit for its Distribution. On the other hand, if the system switched to digital, the VCR would be out of place, provided the available bandwidth accommodated video streams.

TIVO VERSUS VCR: THE PACKAGED PAYLOAD

The VCR was designed and optimized for analog video. It didn't require any format conversion or content processing. As we discussed earlier, video packaged as tapes was easy to rent and share. Should the system switch to digital streaming, the digital video recorder, either virtual or physical, would have a large advantage. Because digital content would be easy to partition and manipulate, the TiVo functionality had the potential to help create highly customized shows with favorite episodes, new audio and video mixes, relevant ad insertion, social interactions, and so on.

The mismatch with the Packaged Payload was so fundamental in the TiVo versus VCR tech battle that a simple switch to digital content would not help the TiVo. For example, having movies packaged in digital form, such as DVDs, didn't provide any advantages to the device because most interfaces between other elements in the video system (in the home or the majority of broadcasts) were based on analog streams.

TIVO VERSUS VCR: THE CONTROL

Control was one functional area where TiVo had considerable advantages. For example, the device could request and receive detailed EPG data from the Internet. It also made it easy for users to express their content preferences, provided advanced content navigation and time-shifting services, and helped personalize content stored locally on the device. With regard to the Control, the inventors of TiVo solved a Synthesis problem. They came up with a new way to interact with video streams. No VCR, whether it could extract the EPG data from cable broadcast or not, was capable of doing that. Furthermore, any other kind of digital aboutness could be integrated into the system, including information about other users' preferences, purchasing decisions, social interactions, and much more.

When we consider the Control as a subsystem, its weakest element was the distribution. That is, the dial-up connection with the sources of EPG was brittle. It was physically inconvenient (remember the long phone cable) and required a paid subscription on top of the already high price for the device itself. Despite that, the Control was an area with the greatest potential advantage because it had an entire S curve ahead of it.

From the perspective of an inventor or a long-term patent portfolio manager, the situation was perfect because the device was far ahead of its competition. As an inventor, the company had a wide-open area to develop their ideas. The tailwinds—the growth of the Internet, the emergence of social networking, proliferation of new video formats for stationary and mobile devices, the development of wireless networking in the home—were all blowing in the right direction.

But from a short-term market point of view, the device itself was a misfit. Enthusiasts and many early adopters loved it, but for the early majority, the headwinds of price and various setup issues were too much to overlook. In 1999, at the height of the dot-com stock bubble, the company went public in anticipation of never-ending Internet money. A year later it crashed with the rest of the stock market, when expectations of large cash infusions evaporated. Over time, the company's financial fortunes fluctuated according to the outcomes of its patent lawsuits against various set-top box (STB) manufacturers who implemented TiVo-like functionality inside their boxes. While the invention had won, the device box had not.

CONCLUSIONS REGARDING THE TECH BATTLE TECHNIQUE

The Tech Battle is an example of an innovation analysis method designed using the three core approaches covered in the book: the system model, multilevel thinking, and S-curve analysis.

The first purpose of the technique is to separate hype and popular misconceptions from reality by elevating the focus of our attention above the solutions' specific features and considering how they fit into the existing or emerging systems. Another purpose is to discover opportunities for short- and long-term innovation by identifying new possible system configurations and element implementations. In addition to that, the Tech Battle helps us understand and appreciate the fundamental difference between linear calendar (astronomical) time used to schedule routine processes and highly nonlinear timing of events within innovation processes. For better results, we recommend using the technique together with the scale-time-money operator and the Climb on the Roof tool from the Three Magicians method.

REVIEW QUESTIONS

1. Use the Tech Battle technique to compare Google Glass to iPhone. Which device has the advantage in the short term and long term? Explain your reasoning.
2. Use the Tech Battle technique to compare the self-driving (robotic) vehicle to the "normal" (circa 2013) vehicle. Identify areas of innovation that would enable robotic vehicles to dominate the landscape. Compare the situation with that for air vehicles.
3. According to the *New York Times* [110]:

> The Energy Department will establish a research hub for batteries and energy storage at Argonne National Laboratory in Lemont, Ill., and spend up to $120 million over the next five years, the department announced on Friday.
>
> Eric D. Isaacs, the director of Argonne, said, "… the hub is looking for the next big thing, something that will store five to 10 times as much energy in a package of the same size and weight."

Assuming the scientists succeed in creating a breakthrough in energy storage, what impact will it have on systems that transfer mass, energy, and information? Sketch out specific system implementations. For example, how will the new technology affect the electric energy grid? Which system element(s) is it going to affect the most: the Tool, Source, Distribution, Packaged Payload, or Control? Explain your reasoning using the Tech Battle technique.

4. Assuming that Google Fiber represents a Distribution Build-up phase, invent or outline major features of a matching Control system. What would it take to make the system a successful innovation?

REFERENCES SECTION III

1. Shockley, William R., The path to the conception of the junction transistor. *IEEE Transactions in Electronic Devices* 23, 7, July 1976.
2. Schumpeter, Joseph A., *Capitalism, Socialism, and Democracy*. 3rd Ed. (New York: Harper and Brothers, 1950).
3. Burgelman, Robert A., Clayton M. Christensen, and Steven C. Wheelwright, *Strategic Management of Technology and Innovation*. 5th Ed. (New York: McGraw-Hill Irwin, 2009).
4. West, Geoffrey, TED Talk presentation. Edinburgh, UK. July 2011.
5. Rogers, Everett M., The diffusion of home computers among households in Silicon Valley, *Marriage & Family Review* 8, 1–2 (1985): 89–101, http://dx.doi.org/10.1300/J002v08n01_07
6. Pervoe rasporyazhenie mera Sobyanina. *Lifenews*. October 22, 2010. http://lifenews.ru/news/41493 (in Russian).
7. Rogers, Everett M., *Diffusion of Innovations* (NewYork: Free Press, 2003).
8. Rubincam, David P., Electronic book. US Patent 4159417, filed Oct. 28, 1977, and issued Jun. 26, 1979.
9. Levy, Steven, *In the Plex: How Google Thinks, Works, and Shapes Our Lives*. (New York: Simon & Shuster, 2009).
10. Heilmann, John, Reinventing the wheel. *Time Magazine*. December 2, 2001. http://www.time.com/time/business/article/0,8599,186660,00.html
11. James, Philip T., Rachel Leach, Eleni Kalamara, and Maryam Syayeghi, The worldwide obesity epidemic, *Obesity Research* 9, S11 (2012): 228S–233S, doi: 10.1038/oby.2001.123, http://onlineli- brary.wiley.com/doi/10.1038/oby.2001.123/full#f1
12. Cartlidge, Edwin, "Tantalizing" hints of room-temperature superconductivity, *Nature*. September 18, 2012. http://www.nature.com/news/tantalizing-hints-of-room-temperature-superconductivity-1.1143
13. Amazon says e-books now top hardcover sales. *The New York Times*, July 19, 2010. http://www.nytimes.com/2010/07/20/technology/20kindle.html?_r=0
14. Ludwig, Sean, Study: Google+ and Facebook use same ad-placement playbook. *VentureBeat*. August 16, 2011. http://venturebeat.com/2011/08/16/study-google-and-facebook-using-same-ad-placement-playbook/

15. Chao, Loretta, Copycat Apple store prompts China investigation, *The Wall Street Journal*. July 25, 2011. http://online.wsj.com/article/SB10001424053111904772304576465763611866934.html

16. Teaching with documents: Eli Whitney's patent for the cotton gin. National Archives. http://www.archives.gov/education/lessons/cotton-gin-patent/

17. Suárez, Fernando F. and James M. Utterback, Dominant designs and the survival of firms, *Strategic Management Journal* 16 (1995): 415–430, doi: 10.1002/smj.4250160602

18. Wozniak, Steve, *iWoz: Computer Geek to Cult Icon* (New York: W.W. Norton. 20036).

19. Tversky, Amos and Daniel Kahneman, Availability: A heuristic for judging frequency and probability, *Cognitive Psychology* 5 (1973): 207-232.

20. Buy or Sell-AMSC: Caught in the crosswinds. Reuters (UK edition), June 9, 2011. http://uk.reuters.com/article/2011/06/09/buysell-amsc-idUKL3E7H62AE20110609

21. AMSC Conference Call with Investors. September 15, 2011.

22. Reich, Leonard S., *The Making of American Industrial Research: Science and Business at GE and Bell, 1876-1926: Studies in Economic History and Policy: USA in the Twentieth Century* (Cambridge, UK: Cambridge University Press, 1985).

23. Moritz, Scott and Olga Kharif, Sprint CEO Hesse seeking more deals as data demand surges. *Bloomberg News*, February 20, 2013. http://www.bloomberg.com/news/2013-02-20/sprint-ceo-hesse-seeking-more-deals-as-data-demand-surges-tech.html

24. *The Economist*. June 2-9, 2012.

25. Goldman, Sylvan N., Combination Basket and Carriage. US Patent 2,155,896, filed May 4, 1937, and issued April 25, 1939.

26. Goldman, Sylvan N., Folding Basket Carriage For Self-Serving Stores. US Patent 2,196,914, filed March 14, 1938, and issued April 9, 1940.

27. Industrial buildings across US to go solar. CNet, June 23, 2011. http://news.cnet.com/8301-11128_3-20073713-54/industrial-buildings-across-u.s-to-go-solar/

28. Google to government: Let us build a faster Net. CNet, September 27, 2011. http://news.cnet.com/8301-30685_3-20112042-264/google-to-government-let-us-build-a-faster-net/

29. Linley, Matthew, U.S. DOE commits $1.4B partial loan guarantee to rooftop solar panel project. *VentureBeat*, June 23, 2011. http://venturebeat.com/2011/06/23/doe-project-amp/

30. Bullis, Kevin, Why Boston Power went to China, *MIT Technology Review*, December 6, 2011. http://www.technologyreview.com/energy/39273/page2/

31. Marshall, Michael, India's panel price crash could spark solar revolution, *New Scientist*, February, 2012. http://www.newscientist.com/article/mg21328505.000-indias-panel-price-crash-could-spark-solar-revolution.html

32. Pearson, Natalie Obiko, Solar cheaper than diesel making India's Mittal believer: Energy, *Bloomberg News*, January 24, 2012. http://www.bloomberg.com/news/2012-01-25/solar-cheaper-than-diesel-making-india-s-mittal-believer-energy.html

33. A construction update, *Google Fiber Blog*, April 04, 2012. http://googlefiberblog.blogspot.com/2012/04/construction-update.html

34. Car balk. Snopes.com/October 14, 2010, http://www.snopes.com/humor/jokes/autos.asp

35. Moore's Law inspires Intel innovation. http://www.intel.com/content/www/us/en/silicon-innovations/moores-law-technology.html

36. Moore, Gordon, Cramming more components onto integrated circuits. *Electronics* 38, 8 (April 19, 1965).

37. Grossman, Lev, 2045: The year man becomes immortal, *Time Magazine*, February 10, 2011. http://www.time.com/time/magazine/article/0,9171,2048299,00.html

38. Keltner, Dacher and Jonathan Haidt, Approaching awe, a moral, spiritual, and aesthetic emotion, *Cognition & Emotion*, 17, 2, (2003): 297–314.

39. Kilby, Jack S., Miniaturized Electronic Circuit. US Patent 3,138,743, filed Feb. 6, 1959, and issued June 23, 1964.

40. Noyce, Robert N., Semiconductor Device-and-Lead Structure. US Patent 2,981,877, filed July 30, 1959, and issued April 25, 1961.

41. The Nobel Prize in Physics 2000: Zhores I. Alferov, Herbert Kroemer, Jack S. Kilby, Nobelprize. Org. press release, October 10, 2000. http://www.nobelprize.org/nobel_prizes/physics/laureates/2000/press.html

42. Gutmann, Ronald J., Innovation in integrated electronics and related technologies: Experiences with industrial-sponsored large-scale multidisciplinary programs and single investor programs in a research university. In *The International Handbook on Innovation*, Larisa V. Shavininia, Ed. (Oxford, UK: Elsevier Science, 2003).

43. *International Technology Roadmap for Semiconductors*. 2007 Edition. http://www.itrs.net/links/2007itrs/execsum2007.pdf

44. Reed, G.T., G. Mashanovich, F.Y. Gardes, and D.J. Thomson, Silicon optical modulators. *Nature Photonics* 4, (2010): 518–526.

45. Alduino, Andrew and Mario Paniccia, Interconnects: Wiring electronics with light. *Nature Photonics* (2007). doi: 10.1038/nphoton.2007.17

46. Baehr-Jones, Tom, Thierry Pinguet, Patrick Lo Guo-Qiang, Steven Danziger, Dennis Prather, and Michael Hochberg, Myths and rumours of silicon photonics. *Nature Photonics*. doi:10.1038/nphoton.2012.66

47. Chen, Xiangyu, Deji Akinwande, Kyeong-Jae Lee, Gael F. Close, Shinichi Yasuda, , Bipul C. Paul, Shinobu Fujita, Jing Kong, and Hon-Sum Philip Wong, Fully integrated graphene and carbon nanotube interconnects for gigahertz high-speed CMOS electronics, *IEEE Transactions on Electron Devices*, 57, no.11, (November 2010): 3137–3143. doi: 10.1109/TED.2010.2069562

48. Gottlieb, Martin and James Glantz, The blackout of 2003: The past; The blackouts of '65 and '77 became defining moments in the City's history *The New York Times*, August 15, 2003. http://www.nytimes.com/2003/08/15/us/blackout-2003-past-blackouts-65-77-became-defining-moments-city-s-history.html?pagewanted=3&src=pm

49. Poulsen, Kevin, Software bug contributed to the blackout, *Security Focus*, February 11, 2004. http://www.securityfocus.com/news/8016

50. Minkel, J.R., The 2003 Northeast blackout—Five years later, *Scientific American*, August 13, 2008. http://www.scientificamerican.com/article.cfm?id=2003-blackout-five-years-later

51. Norman, Donald A., Technology first, needs last: The research–product gulf, *Interactions* 17, 2 (2010): 38–42 doi: 10.1145/1699775.1699784.

52. Berners-Lee, Tim et al., *Weaving the Web: The Original Design and Ultimate Destiny of the World Wide Web by Its Inventor* (San Francisco: HarperInformation, 2000).

53. Berners-Lee, Tim. *Answers for Young People*. http://www.w3.org/People/Berners-Lee/Kids.html

54. Page, Larry and Eric Schmidt, A talk at Stanford's Entrepreneurship Corner. May 01, 2002. http://ecorner.stanford.edu/playerPopup.html?groupId=39

55. Brooks, David, *The Social Animal: The Hidden Sources of Love, Character, and Achievement* (New York: Random House, 2011).

56. Wikipedia. Dunbar Number. http://en.wikipedia.org/wiki/Dunbar%27s_number

57. Takahashi, David, Facebook rejiggers platform to increase game usage, *VentureBeat*, December 22, 1011. http://venturebeat.com/2011/12/22/facebook-rejiggers-platform-to-increase-game-usage/

58. The Web is dead, *Wired*, April, 2010. http://www.wired.com/magazine/2010/08/ff_webrip/all/1

59. Katz, Ian, Interview with Sergey Brin, *The Guardian*. April 15, 2012. http://www.guardian.co.uk/technology/2012/apr/15/web-freedom-threat-google-brin

60. Morgan Stanley Research. February 14, 2011. Tablet Demand and Disruption. Morgan Stanley Blue Paper.

61. Bettencourt, Luís M.A., José Lobo, Dirk Helbing, Kristian Cühnert, and Geoffrey B. West, Growth, innovation, scaling, and the pace of life in cities, *Proceedings of the National Academy of Scientsits* 104, 17 (2007): 1703–1706. www.pnas.org/cgi/doi/10.1073/pnas.0610172104

62. Bergstein, Brian, Mark Pincus on what makes Zynga hum: Short attention spans, *MIT Review*, November 29, 2011, http://www.technologyreview.com/news/426209/mark-pincus-on-what-makes-zynga-hum-short-attention-spans/

63. Dugan, Regina. Presentation at DARPA Cyber Colloquium, YouTube, 2012. http://youtube/Hr0JV0rB4Tg

64. Monthly minutes per active user, photo and video apps, Flurry Analytics. http://blog.flurry.com/Portals/41620/images/Photo&Video_minutesPerMonth_july2011-mar2012-resized-600.png

65. Bilton, Nick, Google begins testing its augmented reality glasses, *The New York Times*, April 4, 2012. http://bits.blogs.nytimes.com/2012/04/04/google-begins-testing-its-augmented-reality-glasses/

66. Baran, Paul, *On Distributed Communications Networks* (Santa Monica, CA: The Rand Corporation, 1962).

67. Steele, Emily and Geoffrey A. Fowler, Facebook in privacy breach, *The Wall Street Journal*. October 17, 2010. http://online.wsj.com/article/SB10001424052702304772804575558484075236968.html

68. Angwin, Julia and Jeremy Singer-Vine, Selling you on Facebook, *The Wall Street Journal*. April 10, 2012. http://online.wsj.com/article/SB10001424052702303302504577327744009046230.html

69. Reynolds, Bernardo, Jayant Venkatanathan, Jorge Gonçalves, and Vassilis Kostakos, Sharing ephemeral information in online social networks: Privacy perceptions and behaviour. In INTERACT '11 *Proceedings of the 13th IFIP TC 13 International Conference of Human–Computer Interaction*, Volume III, pp. 204–215, P. Campos et al. (Eds.) (Berlin: Springer-Verlag, 2011).

70. Largest Data Loss Incidents, http://datalossdb.org/index/largest

71. Giles, Jim, Inside Facebook's massive cyber-security system, *New Scientist*. October 26, 2011. http://www.newscientist.com/article/dn21095-inside-facebooks-massive-cybersecurity-system.html

72. Shteyn, Eugene, The 21st-Century Facebook Utopia, *Eugene Shteyn's Blog*, Google, October 26, 2011. https://plus.google.com/109528994667446704567/posts/iurqshgzm14

73. Singh, Madhur, Angry Mumbai wants answers, changes. *Time*, December 1, 2008. http://www.time.com/time/world/article/0,8599,1862893,00.html

74. What is 3D printing? An Overview. http://www.3dprinter.net/reference/what-is-3d-printing

75. Huang, Cary, The Scale of the Universe-2, 2012. http://htwins.net/scale2/

76. *The Economist*. Prophets of Zoom. Technology Quarterly: Q2, 2012. http://www.economist.com/node/21556097

77. Young, Patrick. TV schedule system and process. US Patent 4,706,121, filed May 6, 1986, and issued Nov. 10, 1987.

78. Shteyn, Y. Eugene. Time- and location-driven personalized TV. US Patent 6,611,654, filed April 1, 1999, and issued August 26, 2003.

79. Shteyn, Y. Eugene. PVR set-up over the Internet. US Patent 7,403,693, filed June 25, 2003, and issued July 22, 2008.

80. Hallet, Anthony and Diane Hallet, *Entrepreneur Magazine Encyclopedia of Entrepreneurs*. (New York: John Wiley & Sons, 1997).

81. Mayo, Anthony J. and Niltin Nohria, The truck driver who reinvented shipping, October 5, 2005. Harvard Business School. http://hbswk.hbs.edu/item/5026.html

82. Cudahy, Brian, *Box Boats: How Container Ships Changed the World* (Bronx, NY: Fordham University Press, 2006).

83. Cudahy, Brian J., The containership revolution. *TR News*. September-October 2006, 246: 5–9. (Washington, DC: Transportation Research Board of the National Academies).

84. *The New York Times*. August 18, 1853. Railroad disasters—Use of the Telegraph. Letter to the Editor of the New York Daily Times.

85. Stephenson, Neal, *Anathem* (New York: William Morrow, 2008).

86. Ali, Kamal et al., Intelligent system and methods. US Patent 7,840,986, filed Dec 14, 2000, and issued Nov 23, 2010.

87. Duhigg, Charles, *The Power of Habit: Why We Do What We Do in Life and Business* (New York: Random House, 2008).

88. Kahn, David. The history of steganography. *Lecture Notes in Computer Science* 1174 (1996): 1–5. doi: 10.1007/3-540-61996-8_27.

89. The Men Who Cracked Enigma. 2004. http://www.imdb.com/title/tt1157073/ https://www.youtube.com/watch?v=eoK4i0SU3DA

90. Sachtman, Noah, Exclusive: Google, CIA invest in "Future" of Web monitoring, *Wired*, July 28, 2010. http://www.wired.com/dangerroom/2010/07/exclusive-google-cia

91. Johansson, Fredrik et. al. Detecting emergent conflicts through Web mining and visualization, Intelligence and Security Informatics Conference, 2011.

92. Adner, Ron and Daniel Levinthal, Demand heterogeneity and technology evolution: Implications for product and process innovation. *Management Science* 47, 5 (2001): 611–628.

93. Hindo, Brian and Brian Grow, Six Sigma: So Yesterday? *Bloomberg Businessweek*, June 10, 2007. http://www.businessweek.com/stories/2007-06-10/six-sigma-so-yesterday

94. Hartman, Peri et al., Method and system for placing a purchase order via a communications network. US Patent 5,960,411, filed Sept 12, 1997, and issued Sept 28, 1999.

95. Systrom, Kevin and Mike Krieger, Stanford University Entrepreneur's Corner Presentation. http://ecorner.stanford.edu/authorMaterialInfo.html?mid=2735

96. Estrin, Judy L. Closing the Innovation Gap. http://sandhill.com/article/closing-the-innovation-gap/

97. Markoff, John, Apple's visionary redefined digital age, *The New York Times*, October 5, 2011. http://www.nytimes.com/2011/10/06/business/steve-jobs-of-apple-dies-at-56.html?_r=3&pagewanted=2&hp/

98. Christensen, Clayton, *The Innovator's Dilemma* (New York: HarperBusiness, 2000).

99. Blank, Steven Gary, *The Four Steps to Epiphany*, Cafepress.com, 2005. http://www.amazon.com/The-Four-Steps-Epiphany-Successful/dp/0976470705

100. Gladwell, Malcolm, The ketchup conumdrum. *The New Yorker*, September 6, 2004. http://www.gladwell.com/2004/2004_09_06_a_ketchup.html

101. Kvamme, E. Floyd, Life in the Silicon Valley: A first-hand view of the region's growth, in Lee, C.M. et al. (Eds.) *The Silicon Valley Edge: A Habitat for Innovation and Entrepreneurship*. (Palo Alto, CA:Stanford University Press, 2000).

102. Gundotra, Vic. Icon ambulance, Google+, August 24, 2011. https://plus.google.com/107117483540235115863/posts/gcSStkKxXTw/

103. Murph, Darren, Apple Q2 2012 earnings: $39.2 billion in revenue, net profit of $11.6 billion, *EngGadget*, April 24, 2012. http://www.engadget.com/2012/04/24/apple-q2-2012-earnings-report-ipad-iphone-sales/

104. Foresman, Chris, iOF app success is a "lottery": 60% or more of developers don't break even, *ArsTechnica*, May 2012. http://arstechnica.com/apple/2012/05/ios-app-success-is-a-lottery-and-60-of-developers-dont-break-even/

105. Murph, Darren, Flurry's analytics: Apple's App Store revenue still leading, but Amazon Appstore close behind, *EngGadget*. March 31, 2012. http://www.engadget.com/2012/03/31/flurrys-analytics-apple-app-store-amazon-app-store-android-google-play/

106. Hu, Jim, Yahoo sheds Inktomi for new search technology. c|net, June 26, 2000. http://news.cnet.com/2100-1023-242392.html

107. Buffet, Warren. 1977 Letter to shareholders. http://www.berkshirehathaway.com/letters/1977.html

108. Diamond, Jared, *Guns, Germs, and Steel* (New York: Norton, 2005).

109. Shteyn, Y. Eugene et al. US Patents 7,529,806; 7,403,693; 7,620,703; 6,918,123; 6,611,654; US Patent Applications 20020116471; 20020078006; 20020162109; 20030037139; 20030069964

110. Wald, Matthew L., Seeking to start a Silicon Valley for battery science, *New York Times*, November 30, 2012, http://green.blogs.nytimes.com/2012/11/30/seeking-to-start-a-silicon-valley-for-battery-science/

Index

Lower case n following page numbers indicates footnotes

C

J

Q

R

S

U

V

Y

Z